AF553942

TOURISM AND ECONOMIC DEVELOPMENT

Valuing Environmental Demand for Chilika Lake Through Travel Cost and Contingent Valuation Methods

TOURISM
AND
ECONOMIC DEVELOPMENT

Valuing Environmental Demand for Chilika Lake Through Travel Cost and Contingent Valuation Methods

By

Deepak Bishoyi

DISCOVERY PUBLISHING HOUSE

NEW DELHI-110002

First Published-2007
Reprint : 2012
ISBN 978-81-8356-193-8

© Author

Published by

DISCOVERY PUBLISHING HOUSE
4831/24, Ansari Road, Prahlad Street,
Darya Ganj, New Delhi-110002 (India)
Phone: 23279245 • Fax: 91-11-23253475
E-mail: dphbooks@rediffmail.com
dphtemp@indiatimes.com

Printed at:
Dynamic Printers

Preface

Tourism is one of the valuable attributes to the economic growth of most of the developing countries. It can act as the pivot of vehicle for economic development. Today, tourism has been acknowledged as the biggest component of the tertiary sector. It constitutes a major portion in the world trade, and its growth rate is faster than that of the tangible goods. Tourism has become a highly organized industry consisting 11 per cent of World's Gross Product. World's revenue from tourism has increased from $70 billion in 1960 to $1500 billion in 1995. Since time immemorial India has been a great source of attraction to the world, but tourism industry has not developed so far to its full potential. It contributes only 0.3 per cent to world tourism. India's share to the global tourism industry is quite insufficient considering the continental size of the country and its rich tourist resources. Therefore, renewed efforts are made to catch up with the rest of the world in this field.

The importance of tourism as a contributor of economic growth is widely accepted year after year throughout the world. Now however, as an account of fast expansion of tourism, we can cite a large number of economic benefits flowing from this industry. The major ones are:

I. Tourism is now a part and parcel of International Trade in services, in fact occupying the top position;

II. It is a significant component of national economies;

III. It is a means of earning foreign exchange;

IV. It is a provider of employment;

V. It is a powerful factor of development.

First we take up the issue of contribution of tourism to the national income of a host country. Tourists have to pay for different types of services and goods in the host country. The growth rate of tourism is faster than any other trade in the world. The percentage of earnings from tourism in the total exports of many countries has been increasing year after year. It has been universally accepted that if a country is able to earn foreign exchange of a minimum of 10 per cent of the merchandise exports from tourism, the country can be called a 'Tourism' country. It is abundantly clear that tourism in India is significantly contributing towards strengthening our balance of payment position. This could be considered as one of the biggest economic gains from tourism. We have now come to an era, which is of immense value to us. Tourism occupies a predominant place in the service industry that can meet the demand for employment of a huge mass.

There are also some negative aspects of Tourism. These are, however, marginal and relate primarily to overcrowding of a tourist spot, natural or man-made. However, even these marginal negative factors must be overcome by a systematic tourism promotion policy. It will be necessary to undertake research works and in-depth studies in different areas before policy formulation takes place. The tourists should be encouraged to visit certain tourist spots under a 'guided' programme. Management of historic sites, monuments, wildlife reserves, etc. should be improved. A mass education programme for the operators of the tourism *plant* of India must be undertaken. The tourist should also be oriented about a particular tourist spot before the actual visit takes place.

Environmental resources are key attractions of tourist demand and constitute the most obvious area in which developing countries do not achieve an optimal return. In some countries, under-utilization of available resources means that returns that might otherwise be available are foregone. In others, excessive use resulting from the public good and under-priced nature of the assets gives rise to a level of degradation, which threatens tourism's resource base and the long-run returns from it. Debates within environmental economics concerning appropriate definitions of sustainability, methods for valuing resources and price or regulation-based techniques for achieving sustainable resource use have indicated possible means of attaining ongoing returns from environmental assets. However, in practice, lack of consensus over the appropriateness of the weak or strong definition of sustainability is reflected in disagreements. Moreover, many developing countries are understandably reluctant to engage in a level of environmental protection that was not observed by industrialized countries during a similar stage of development. The root causes of tourism development that is unsustainable in terms of the stock of resources are to be found in market failure. Many natural resources are public goods, and free access to them often results in over-use. Negative externalities occur, as for example, hotel construction creates additional pollution. One of the main obstacles to achieve the sustainable use of environmental resources is that they are freely available or under-priced. It is, therefore, necessary to attribute a social value to them. The total economic value of the resources consists of use value (which is likely to exceed the price owing to the existence of consumer's surplus), and option value (which takes account of consumer's willingness to pay to ensure the continued existence of the resources, irrespective of whether they will use them).

The work done in this book is related to the need and importance of tourism on economic development and tourism potential at Chilika Lake, Orissa, India. So the study mainly deals with the influence of socio-economic variables on demand for recreation in Chilika, to appreciate the wiilingness to pay of tourists for conservation and protection of Chilika, and estimate the total recreation value of the lake. This book is the outcome of the dissertation submitted to Berhampur University in partial fulfillment of requirements for the degree of Master of Philosophy. It is hoped that the book will be immensely useful in guiding the policy makers, planners, administrators, environmentalist and researchers in formulating policy paradigms in India.

My greatest personal debt is to my Guide, Dr. (Mrs.) Amita Kumari Choudhury, Reader, Department of Economics, Berhampur University, Orissa, who conceived, detailed and shaped the problem and provided sagacious guidance throughout the present study. Her scholarly suggestions, prudent admonitions, immense interest and affectionate behaviour have been a source of great inspiration to me

My highest regards are acknowledged to my family members especially my father whose blessing, moral support and sound advice are responsible for my progress in this study and throughout my academic career. My hearty regards are to Dr. S.N. Tripathy, Reader in Economics, Aska Science College, Aska, Mr. M.N. Pattanayak, Lecturer in Economics, Kandupadar College, and Mr. K.C. Pradhan, Lecturer in Economics, Khemundi College, Digapahandi. Without their blessings and moral support, I may not be able to carryout the present work. I am also thankful to Dr. C. Nanjundaiah, faculty at ISEC, for his intellectual suggestions. I am grateful to all the friends in Berhampur University, Orissa, ISEC Bangalore and IIPS Mumbai for providing me all moral support to finalise this work.

Last, but not the least, is the commendable support provided by Sri Tilak Wasan, Proprietor and other associated personnel of Discovery Publishing House, New Delhi, for having agreed to publish this volume and giving it the shape in which it is placed for scrutiny of the academia, administrators and others interested in the field.

Deepak Bishoyi
Institute for Social and Economic Change (ISEC)
Bangalore

Contents

1

The Problem of the Study

INTRODUCTION

Tourism involves with different sectors of the economy. Properly managed, it confers benefits in terms of social, psychological and economic well being. Travel and tour enhance quality of human experience spiritually and materially (Krippendorf, 1987). It creates employment opportunities, generates revenue, augments foreign exchange earnings, and promotes national integration, and international understanding. It is a smokeless industry and one of the fast growing business world over. Through tourist flow to various countries one can foresee a process of cultural assimilation and economic development. Different economies have variety of benefits to offer to a tourist. It may be in the form of natural scenic spots or man-made wonders like Disneyland, USA. According to World Travel and Tourism Council (WTTC), tourism industry is expected to play a key role in the rejuvenation of the world economy. World travel and tourism business is likely to increase from $3 trillion in 1995 to $7.2 in 2005. During the same period, tourism's share in consumer expenditure is expected to increase from 11.4 per cent to 11.6 per cent, share in capital investment from 11.3 per cent to 11.8 per cent and jobs from 212 million to 312 million.

In the panorama of Indian culture and tourism, Orissa occupies an enviable position being the paradise of the tourists. The state unfolds itself in the beauty and elegance of the Sun Temple of Konark, the Rajarani Temple, the Lingaraj Temple, Mukteswar Temple of Bhubaneswar, and The Jagannath Temple of Puri, the scintillating golden beach of Puri and Gopalpur, the fauna and fountain of Mayurbhanja, Phulbani and Koraput districts.

This chapter is an introductory one. The research design is the core part of this chapter. It contains literature survey on tourism and valuation of natural resources, and the lake Chilika. The objectives are clearly stated. Hypotheses are framed. Methodologies are clearly spelt out.

SCOPE

Chilika is the largest brackish water lagoon in Asia located in the state of Orissa. It is between the latitudes 19° 28′ and 19° 54′ N and longitude 85° 38′ E. The mean widths during the summer and monsoon are 14.07 km and 18.0 km., respectively. The water-spread area of the lake varies between 1165 to 906 sq. km. during monsoon and summer. The lagoon spreads across Ganjam, Puri and Khurda districts of Orissa. The scenic beauty of the Chilika lagoon attracting the tourists mainly because it lies between green hills in south, Bay of Bengal in northeast and numerous aesthetic islands in the East. The lake has a major link with the Bay of Bengal on its southern end through an irregular 29 km. long channel (starting from Satapada) with several small sandy and usually ephemeral islands. About one km. wide, the channel runs parallel to the sea. It is separated from the sea by a very narrow channel known as Magarmukha Muhan. The lake has another link at its southern end through Palur canal and is separated from Chilika by low lands.

It has a unique floral composition. It is home to some rare, vulnerable and endangered species of animals. Total numbers of fish species vary from 149 to 225. There are varieties of phytoplankton, angae and aquatic plants. The lake is the habitat of number of resident and migratory birds. The avifauna of the lake is a source of thrill to the viewers and provide a laboratory for research. The lake attracts as many as 0.13 million domestic visitors and about 250 foreign tourists, during one year. The lake meets the different requirements of about 1 lakh people in 192 villages around the lake. About 50 thousand fishermen derive their sustenance from the lake. It has a significant contribution in terms of revenue earning from tourism. The study is a modest attempt to estimate the recreation value of this precious wetland.

REVIEW OF LITERATURE

There is a growing body of literature that focus on economics of tourism. From three angles literatures are reviewed here. First attempt is made to review literature economics of tourism. Secondly we review the works on Chilika lake. Thirdly literature on valuation of environmental resources is reviewed.

Literature on Economics of Tourism

Mostly the literature which focuses on social and economic effects of tourism development revolves around the debate of positive and negative impact of tourism (Nesh, 1981). The positive effects are increased total incomes and employment generation, and the stimulation of secondary economic growth, educational, scientific and aesthetic advantages (Mc Kean 1977, Boissevain 1977, DeKalt 1979).

Studies to estimate tourism demand have used single equation models and at the aggregate, cross-country level (Archer, 1976; Johnson and Ashworth, 1990; Sheldon, 1990; Sinclair, 1991). Equation (1) exemplifies the approach, regressing tourism demand, D_{ij}, on income per capita. Y_i, relative prices, $P_{ij\,k}$, exchange rates, $E_{ij\,k}$, transport costs, $T_{ij\,k}$ and dummy variables for one-off events, DV.

$$D_{ij} = f(Y_i, P_{ij\,k}, E_{ij\,k}, T_{ij\,k}, DV) \quad (1)$$

where i refers to the tourist origin country, j refers to the destination and k to competing destinations.

One of the interesting feature is the variation in the estimated values. The study by Gray (1966), provided per capita income elasticities of 5.13 for US demand for tourism in the rest of the world and 6.6 for Canadian demand. Jud and Joseph (1974) estimated a per capita income elasticity of 1.74 for US expenditure on tourism in 17 Latin American countries, and of 2.58 for total international demand. Stronge and Redman (1982) estimated an income elasticity of 2.99 for US expenditure on Mexican tourism. The per capita income elasticities for expenditure in Turkey by 11 industrialised countries ranged between 0.92 and 5.95 (Uysal and Crompton, 1984). A set of income elasticities has also been estimated for tourist arrivals and includes values of 3.15 for US arrivals in Hawaii (Bechdolt, 1973), 1.59 to 6.07 for arrivals in Turkey from a range or origins (Uysal and Crompton, 1984), 0.81 to 7.3 for Singapore (Gunadhi and Boey, 1986), 0.18 to 8.1 for Fiji (Broomfield, 1991), 0.21 for Reunion (Seeberger, 1992) and 0.94 to 3.44 for Malaysia (Shamsuddin, 1995).

First we can take a critical review of these models. The single equation approach to tourism demand modelling lacks an explicit theoretical basis and takes no account of the micro foundations of demand (Sinclair and Stabler, 1997). The nature of consumer decision-making concerning expenditure on tourism relative to other goods and services, and the allocation of time to tourism relative to paid work or unpaid activities, are not examined. Second, all of the studies have ignored possible inter-temporal relationships between tourism expenditure and income or relative prices/exchange rates, (Deaton, 1992). Third, little attention has been paid to the issue of inclusion of relative prices and exchange rates as separate determinants of demand. Fourth, there is doubt regarding the cost of transport between the origin and destination in empirical works. Fifth, most studies fail to include the full range of test statistics.

The econometric model of tourism expenditure in the south Mediterranean destinations of Greece, Italy, Portugal, Spain and Turkey estimated by Syriopoulos (1995) tackles the second, third and fifth problems by estimating a dynamic model of demand. This approach has the advantage of separating short-run and long-run elasticities.

System of equations models of tourism expenditure allocation posit that consumers make decisions according to a stage budgeting process. The consumer's decision-making initially involves allocating the expenditure budget among broad groups of goods and services, for example, food, housing and tourism, and subsequently among sub-groups such as holidays in Africa, Asia, Europe, North America and other regions of the world. Expenditure is then allocated to countries within the chosen regions and to types of tourism expenditure within the chosen countries. The approach has been used to explain the allocation of the tourism expenditure budget among different countries (White, 1982; O'Hagan and Harrison, 1984; Smeral, 1988; Syriopoulos and Sinclair, 1994) and among types of tourism expenditure (Fujii et al., 1987; Sakai, 1988; Pyo et al., 1991).

The Almost Ideal Demand System (AIDS) model formulated by Deaton and Muellbauer (1980a. 1980b) provides an attractive specification for tourism demand estimation, incorporating both the axioms of consumer choice and the stage budgeting process. Its main concern is to explain changes in the budget shares of tourism expenditure attributed to destinations or goods and services rather than changes in levels of tourism demand which are the focus of single equation models. The model incorporates PIGLOG (price-independent generalised linearity) consumer preferences which allow for perfect aggregation over consumers and uses the cost (expenditure) function *e(u.p.)* that defines the minimum expenditure necessary for a level of utility. *U.* given prices. *P*. This may be written as:

$$\text{Log}\ e(u, p) = (1-u) \log \{a(p)\} + u \log \{b(p)\}$$

Linear, homogeneous. Concave functions of prices, *a(p)* and *b(p)* apply to the price vector, *p,* and the functional forms used are:

$$\log \quad \alpha(p) = \alpha_o + \sum_{i=1}^{n} \alpha_i \log p_i + \frac{1}{2} \sum_{i=1}^{n} \sum_{j=1}^{n} \gamma_{ij}^{*} \log p_i \log p_j$$

and

$$\log b(p) = \log \alpha(p) + b_0 \prod_{i=1}^{n} p_{p_i}^{b_i}$$

The AIDS cost (expenditure) function is given by

$$\log c(u, p) = \alpha_o + \sum_{i=1}^{n} \alpha_i \log p_i + \frac{1}{2} \sum_{i=1}^{n} \sum_{j=1}^{n}$$

$$\gamma_{ij}^{*} \log p_i \log p_j + ub_o \prod_{i=1}^{n} p_{p_i}^{b_i} \qquad (2)$$

where a_1, b_1 and γ_{ij}^* parameters.

Since the price derivatives of the cost function are the quantities demanded

$$\frac{\partial c(u,p)}{\partial p_i} = q^i \tag{3}$$

If both sides of equation (3) are multiplied by $p_i/c(u. p)$ we obtain, in log form

$$\frac{\partial \log c(u,p)}{\partial \log p_i} = \frac{p_i q_i}{c(u,p)} = w_i \tag{4}$$

where w_1 is the budget share of the *l*th good. Using equation (2) and differentiating with respect to log p_i using equations (3) and (4), and letting $\gamma_{ij} = \frac{1}{2}\left(\gamma_{ij}^* + \gamma_{ij}^*\right)$,the IDS model is obtained

$$w_i = \alpha_1 + \sum_{j=1}^{n} \gamma_{ij} \log p_j + b_i \log\left(\frac{x}{p}\right) + u_i i = 1......, n \tag{5}$$

Equation (5) defines a system of budget equations for each of n goods, where w_i is the budget share of the ith good, p_i is the price of ith good, x is total expenditure on all goods in the group (system). P is the aggregate price index for the group and u_i is the normal disturbance term with zero mean and constant variance. In the case of the allocation of tourism expenditure among a range of destinations, for example, w_i is the share of the budget that residents of origin j allocate to tourism in destination I, p_j is the effective price level (adjusted by exchange rates) in origin j, x is the budget for tourism expenditure by residents of origin j and p is an index of effective prices in the destinations. Thus, the model takes account of the role of the expenditure budget and relative prices in explaining tourism demand, in accordance with consumer demand theory.

The system of equations approach overcomes a number of the problems which characterise many single equation studies of tourism demand. In particular, it has an explicit theoretical basis which permits aggregation from the individual to all tourism consumers. Progress has also been made towards incorporating inter temporal relationships in demand modelling (Blundell. 1991).

In their work Ehrlich and Ehrlich 1981; Wilson 1985, Lawton and Hay 1995 have pointed out that human activities have contributed to an increase in species extinction. The studies suggested for implementation of safeguarding strategies. Gossling (1999) argues that ecotourism can contribute to safeguard biodiversity and ecosystem functions in developing countries. He argued that tourism and its high direct use value can play an important role as an incentive for protection. He suggested for introduction of the concept of environmental Damage costs and integration of it into the calculation. Further he has analysed the international tourism development and related it to protection goals. Fankhaser (1994) has estimated the social costs of Co_2 emissions in the order of \$ 20/tc for the period 1991-2000. This implies that every tourist travelling by air creates \$ 3.5 of environmental damage. But this estimation is conservative, as air traffic causes additional trace gas emissions (NO_x, SO_2, Co, HC, H_2O), which have a substantial global warming potential, or transformed in physico-chemical processes, severe impacts on ecosystems (Fabian and Karcher, 1997). Kuss et al., (1990) have highlighted the complex impact of tourism on the environment, society and economy. The IUCN (1992) lists tourism as the second major threat to protected areas.

A good number research works are on the relationship between tourism and environment, and problems associated with tourism expansion (Pearce 1985, Romeril 1989, Farrella'nd Runyan 1991, Cater and Goodall 1992, Eber 1992, and Jenner and Smith 1992). It is also recognised that tourism can benefit poor countries by providing a return from their environmental resources (Boo 1990). One methodology which would seem to be appropriate for taking account of the complete set of costs and benefits associated with tourism is cost-benefit analysis (Bryden, 1973). However, application of this technique has generally failed to take account of environmental resources, owing to the difficulties of valuing them and of quantifying the inter-and intra-generational effects of alternative rates and types of utilisation. Studies by Eagles (1992), Blamey (1995), Lang et al (1996) and other demonstrated that ecotourism is a specially product appealing largely to an up market. It has been the most rapidly growing sector of tourism industry (Glannecchini, 1993).

Studies on Lake Chilika

In their work Kadekodi and Nayampalli (2005) presents an ecology-economy interactive simulation model. They have monitored lake sustainability indicators and identified possible policy interventions. It is suggested to treat the problem of biodiversity loss as one of human rights for livelihood and population management through which the trend in biodiversity decline can be reversed.

Choudhury et al. (2005), in their work considers the effect of salinity change and open access rule on the prawn (including shrimp) fishery resources of the Chilika lake, Orissa. The paper considers how the detoriation in the environmental condition and the open access regime have caused the decline in the shrimp catch.

Chauhan (2002) has worked on the "capturing eco-tourism potential at Chilika lake" in Orissa. Chilika is one of the six wetlands of international importance designated by Ramsar convention. Chilika Development Authority was set up in 1992, to promote the lake eco-system. The vast stretches of virgin blue waters and beaches bordering island in the southern sector, the brilliant miracle of a million migratory winter birds, the rare irrawadly dolphins offer ample scope for developing regulated eco-tourism along with potential to generate employment opportunities.

Choudhury (2002) has proposed a community participated eco-tourism programmes that will enhance the livelihood options for the local people and strengthen the bond between nature and man. Keeping the lagoon as the hub of this community based tourism network the proposal intends to link it with Gopalpur beach tourism and a river safari in Rushikulya. Das (2002) throws light role of NGOs local community and international bodies to restore Chilika as an eco-tourism destination.

The paper by Lenka (1998) focuses on community based planning in Chilika and its environment. These are resources use by the community and tourist, preservation of traditional cultural values and conservation of the wetland. Lenka (2002) further emphasised on that the eco-tourism development strategies should be community driven to raise consciousness and awareness of the local people. They can participate with all development agencies. Further this strategy will provide benefit all stakeholders and tourists. Nanda (2002) has emphasized the role of the public sector and private enterprise in creating awarness to promote eco-tourism.

In his work Parida (2002) has reported that community based tourism is a viable option for generating economic opportunities, fostering environment conservation and keeping alive traditional cultures and customs.

Swain (2002)'s study is an attempt to highlight on causes and consequences of environmental degradation in the lagoon. He stressed on the conservation strategies for eco-tourism.

Rath (2000) has measured willingness-to-pay (WTP) of the visitors as an amount over and above the travel cost. The payment vehicle of this WTP is gate of entry fee. Thus the study combines elements of CVM and TCM. Tarai (2003) has estimated the recreational value of Chilika by using Travel Cost Method (TCM). The study has measured the TC and recreational value of Chilika by taking the visitation rate as a dependent variable.

Chase et al., (1998) studied ecotourism demand and the differential pricing of National Park access in Costa Rica. The study presents a conceptual framework and an empirical analysis of the impacts of introducing a differential entrance fee policy at three national parks in Costa Rica.

Literature on Valuation

The root causes of tourism development which is unsustainable in terms of the stock of resources are to be found in market failure. Many natural resources are public goods and free access to them often results in over-use. Negative externalities occur as, for example, hotel construction creates additional pollution. Tourism expansion can also magnify intra-generational inequity as local residents are displaced from their homes as has occurred in, for example, Egypt, the Gambia, Mexico, Morocco, Myanmar and the Philippines (Long, 1991; Tourism Concern, 1995). Further the agricultural and fishing areas are polluted, as in Hawaii (Helu Thaman, 1992).

One of the main obstacles to achieve the sustainable use of environmental resources is that they are freely available or under-priced. It is, therefore, is necessary to attribute a social value to them. The total economic value of the resources consists of use value (which is likely to exceed the price owing to the existence of consumer's surplus), option value which takes account of consumer's willingness to pay to ensure the continued existence of the resources, irrespective of whether they will use them.

There are number of environmental valuation studies but few have used economic approach to calculate welfare measurement. (Bateman and Willis, 1999 and Freeman, 1993). The Lumpinee park study by Grandstaff and Dixon (1986) and TDRI/HIID study on Khau Yai National Park (Kaosaard et al., 1995) are two important studies conducted in Thailand that have used economic valuation methods. Both studies have compiled the travel cost method (TCM) with the open-ended contingent valuation method (CVM) in order to assess willingness to pay (WTP).

A number of techniques have been developed for the purpose of valuing environmental resources, those most usually applied are: the hedonic pricing model (HPM), the travel cost model (TCM) and the contingent valuation model (CVM). The HPM (Rasen, 1974) conceptualises goods as bundles of characteristics and is concerned to estimate the implicit price of each. In the case of tourism, it may be used to estimate the implicit price of environmental resources as well as of the components of tourism, such as accommodation and travel. It has mainly been used to estimate the effect of natural resources such as forestry or waterway proximity on residential property price (Garrod and Willis 1991, Willis and Garrod 1993a) but has also been used to evaluate the location component of package holidays in southern Spain (Sinclair et al. 1990).

The TCM was developed from a suggestion made by Harold Hotelling in 1947 in a release on the economics of recreation in US national park by the National Park Service. Hotelling (1947) suggested measuring differential travel rates according to travel distances that visitors had to overcome in order to reach a park. Exploiting the empirical relationship between increased travel distances and associated declining visitation rates, would permit one to estimate a true demand relationship. If estimated empirically, this demand schedule could be used to compute the total benefits produced to park visitors, which should be equal to any entry fees they paid plus other unpriced benefits or consumer surplus (Hotelling, 1947).

Clawson (1959), Knetsch (1963) and later Clawson and Knetsch (1966) showed how a zonal methodology (ZTCM) could be used to derive a demand curve for a site. The derived demand curve estimated by them appeared generally satisfactory. They exhibited a negative relationship between price and output in accordance with demand theory. Brown and Nawas (1973) and Gum and Martin (1974) developed a new form of TCM based on individual visitors, where the dependant variable, i.e., the quantity consumed, is the number of trips taken per period by individuals or households (ITCM).

The travel-cost model (Clawson and Knetsch. 1966) assumes that the cost of travel to recreational or tourism locations is a measure of consumers' willingness to pay and, hence, of their valuation of the locations and the resources within them. The model has been applied to fishing (Smith et al. 1991) and deer-hunting sites (Loomis et al., 1991). Brown and Henry (1989) applied both the TCM and the contingent valuation model to elephant viewing in Kenya and estimated that it contributed between $23 and $30 million in 1989. The CVM involves asking consumers about their willingness to pay for the resource in question and has also been applied to canals (Wills and Garrod 1993a), forestry (Hanley and Ruffel 1992, 1993), wildlife (Willis and Garrod 1993b) and national parks (Lockwood et al. 1993).

Sutherland and Walsh, (1985) estimated the effect of distance on the preservation value of environmental quality using the CV approach. The case considered is the potential degradation of water quality due to coal mining activity in the Flathead river drainage system, Montana. They made a household survey of WTP for water quality at the Flathead river and lake area.

Li (1994) made a CV survey of forest environment in Vasterbotten in northern Sweden. The main motive of the survey was to assess the monetary value of the forest landscapes as a natural environment for recreational activities given the present forest pattern. He took the sample of 800 individuals, out of which 436 responses were received. After deleting the non-response items, 334 samples were used for estimation. The estimation process is done using the probit model and semi-parametric methods.

Loomis and Caban (1998) conducted a CV survey to estimate the economic value to California and New England residents of implementing a fire management plan to reduce acres of old growth forest that burn in California and Oregon. For statistical estimation, data were collected from 499 California households and 449 New England households through in-depth interview. For estimation purpose they used Probit model.

Pate and Loomis (1997) estimated the influence of distance on the WTP for public goods with large non-use values. The data used to explore this issue came from a CV study completed by (Loomis et al. 1991). Here information were collected from 1003 sample households of San-Joaquin Valley, California residents out side of the valley, Washington State, Oregon state and Nevade state residence. A non-liner probability model is used for estimation. And the result of their study is that distance affected WTP for wetland habitat and wild life and wildlife contamination control programmes. The WTP for these two programmes show a statistically significant negative relations, whereas the Salmon Improvement programme did not.

Katchen and Reiling (2000) tried to measure environmental attributes and estimated non-use values for protection of peregrine falcons and short nose sturgeons, both endangered species in Maine. Information from 1200 samples was collected from the residents of Maine. The respondents were asked to pay one time contribution through increased taxes. For estimation purpose they used the logit regression model. The study shows that people with stronger pro-environmental attitudes places more weight on the importance of ethical reasons for species protection.

Chopra and Kadekodi (1999) applied the CVM for valuation of forests in the Yamuna basin. WTP has been used to assess the user value of a change in the quantity of forest. They took 585 households for sample survey and used the OLS method. Since it shows poor results in terms of R-square and level of significance, they used weighted least square method and maximum likelihood technique. The study reveals that the rural population does place a value on the functions performed and products desired from forests in their vicinity.

Murty (2000) made a CV survey to estimate the benefits from the improvement in the river water quality from 1985 to 1995. A nation wide survey was conducted with 2000 samples. However 1876 filled in questionnaires were entered into spreadsheet software. But the regression analysis was conducted taking 817 samples only. The study shows that income, size of household and educational level are positively related to WTP.

Chopra and Adhikari (2004) have probed the nature of the link in the context of a wetland in Northern India. A dynamic simulation model is developed.

Champ, et al., (2000) investigate the effect of the payment mechanism on contingent values by asking a willingness-to-pay question with one of the three different payment mechanisms: individual contribution, contribution with provision point and referendum.

Alberine, et al., (2003), used the NOAA Panel on CV advocated a 'no answer' response option to dichotomous choice payment questions, but did not give guidance as to how this additional response should be interpreted conceptually or analytically. They investigate the econometric modelling and response effects associated with multiple-bounded, polychotomous-choice payment question. Moreover, in their application, explicitly modelling uncertain responses can increase welfare estimates by over 100 per cent.

Svedsater (2003) has adopted a qualitative approach in order to investigate how people make sense of CV questions of a global environmental amenity. The objective is not only to capture people's motivations and consideration of WTP, and to determine if they adequately comprehend an economic valuation of such public goods. The findings indicate that a large proportion of respondents do not interpret the valuation task, as intended.

Hadker, et al., (1997) have conducted survey in Bombay and elicit their WTP for the maintenance and preservation of Borivili National Park using the CV method. The authors pay special attention to hypothetical bias, starting point bias, embedding effects and part-whole biases.

Raut (2003) conducted a CV survey to estimate the option value of Chandaka forest. Orissa and elicit their WTP to protect Chandaka for future generation. For estimation purpose she has used the probit and logit-models. For this, binary choice variables such as WTPAP and WTPSP are considered as the dependent variables. The estimations of the probit and logit model for both WTPAP and WTPSP using maximum likelihood estimator (MLE) are found quite satisfactory. The mean WTP is estimated from the index function but it gives a higher value than the median WTP under annual payment. A comparison of the WTP of the same in other studies conducted in different parts of the world including India, shows that the figure estimated here is on the lower side and therefore not unreliable.

In their work, Bockstael et al. (1991) refers Travel cost model as continuous demand models where the numbers of trips a recreationist takes in a season depends on price, income, quality of the sites and other socio-economic characteristics. On the contrary McFadden's random utility models (1977) focus on the individual's decision to visit recreation site. The individual's visit to a site that maximises his utility and the probability of choosing a particular site is a function of the difference between the utilities associated with each site choice and their random components.

Seller, Stoll and Chavas (1985) argue that in the travel cost model the price of petrol should be included as cost only. However in many studies apart from the price of petrol, the cost incurred for insurance, depreciation and maintenance have been included in calculating the cost associated with visiting a park.

McConnell (1975) and Cesario (1978) posited that the time taken to travel to the park is an important cost variable in determining demand. If time costs are not included in the TCM, the demand curve will be biased downward (as the travel cost used is less than its true value) and consumer surplus estimates will be conservative (Cesario 1978).

Walsh, Johnson and McKean (1992) conducted a survey of published and unpublished empirical studies in the US. They found that 156 benefit estimates had been completed during the period 1968-88. The purpose of the Walsh, Johnson and McKean study was to analyze empirical results in order to develop an understanding of the factors that are most important in predicting recreational use and benefits to the visitor. They also identified additional explanatory variables including site quality, travel time cost and substitute price.

Ward (1983) and Shaw (1992) concluded that the opportunity cost of an individual's time was not necessarily related to wage rate and information on personal situations and preferences was necessary before appropriate assumptions could be made. McKean, et al. (1995) building on Shaw's (1992) work, theorized that time-rationing rather than time-pricing may be more appropriate given labour market and institutional considerations.

McConnell (1992) concluded that since accounting for on-site is so difficult, no systematic method has been developed, either conceptually or empirically. Freeman (1993) approached the substitute site dilemma by suggesting that researchers ask visitors which other single site is visited frequently and include only that site's price as the relevant substitute price. He asserted that the next-best site yielding similar attributes (a national park, in this instance) is the appropriate alternative.

Most of the CV studies adopt binary choice models. People's WTP may not have systematic relationship with explanatory variables to be captured by OLS. Many of the responses may suffer from several limitations like strategic bias or interviewer bias or truncation problem. As a result what is important to consider is whether a respondent is willing (yes) or not willing to pay. The NOAA Panel (Hanley et al., 1997, Alberini, Boyle and Welsh, 2003) has also prescribed the use of qualitative response (OR) models have been a industry in econometrics, where a number of new techniques have developed (Greene, 2000).

As against this, the present study adopts the TC of the domestic visitors to Chilika and estimate the TC and recreational value by taking the number of visit to Chilika as a dependent variable (Garrod and Willis, 1999). In this sense, the work has currency and worth.

OBJECTIVES

The present study has the following objectives:

(i) To review the national and international literature on economics of tourism and economic valuation of natural resources;

(ii) To determine the influence of socio-economic variables on demand for recreation in Chilika;

(iii) To appreciate the willingness to pay of tourists for conservation and protection of Chilika;

(iv) To estimate the total recreation value of the lake.

METHODOLOGY

The work is based on both secondary as well as primary data. The secondary data includes the relevant literature and background information on Chilika Lake, literature on economic valuation and methods of estimation. The data on the lake are collected from the libraries of Berhampur University, Chilika Development Authority and Orissa Tourism Department (OTDC). Information available from these sections are processed and analysed in Chapter – III. The meaning and methods of estimation of recreational value are developed from the materials available in the libraries of the universities and research centres of India.

Sampling is a critical issue for travel cost studies. While some scholars prefer stratified sampling from the total population (Choe, et. al, 1996; Hanker, et. al, 1997), others use random sampling from user groups only (Farber, 1988; Yaping, 1998). The work concentrated on the user group. Individual visitors instead of households were chosen as respondents for interviews. "Visitors" were broadly defined as those who use the ake for recreation.

Since the population is large, only 110 numbers of respondents was taken as a sample for this study. This study used systematic random sampling where every 10th visitors was interviewed. In case s/he refused, the next visitor was interviewed. The analysis is based on the sample survey conducted by face to face interview with the help of a questionnaire-cum-schedule to the tourists who came from various places of the India. This survey was conducted during the months of December 2004 and January 2005 at different times and different places near Chilika belonging to different age, sex and income group. The primary data enables us to know the opinion of tourists about Chilika lake, tourism provision by the Government of Orissa at Chilika tourists facility at Chilika, and its infrastructure problem.

Primary data is collected by conducting a survey of 110 questionnaires cum schedule by face-to-face interview. The questionnaire is divided into nine sections. The first section is the general information about the tourist, next is the socio-economic information. Visitors recreational particulars and purpose of visit are discussed in the third and fourth sections respectively. The next section describes on the mode of travel to lake, visitor's expenditure on different items etc. Section eight deals with the different parts of lake visited by tourist. The last section is on visitors attitude about the lake.

This survey is conducted during the month of December 2004 and January 2005 at different times and different places near the lake from tourists belonging to different age, sex and income groups. The data collected from the survey are processed and presented in a series of tables in Chapter III. To estimate the recreational value standard econometric tools like regression models, and logit and probit models are used.

HYPOTHESES

The study verifies the following hypotheses:

1. Chilika lake is a unique tourist spot in Orissa;
2. There is an inverse relationship between number of visits to the lake and the cost of travel;
3. It is surmised that tourists willingness to pay for the conservation of the lake tend to increase with rise in income;
4. The total recreational value of Chilika is high enough to justify protection and conservation of the lake.

LIMITATIONS

The study has certain limitations. The first difficulty is that the sample size is very low. A standard CV and TCM study should have covered a far larger samples than what could be possible here. The second weakness relates to the exclusion of substitute sites in our data collection. Our observation would have much better with inclusion of it.

CHAPTER DIVISION

The dissertation is organised in five chapters. First, the introductory chapter contains the review of literature, objectives, scope, methodologies and the hypotheses of the study. Chapter two contains meaning and concept of tourism, economic contribution of tourism, valuation of environmental resources and economics of tourism in Orissa. The tourism potential of the lake is narrated, in the third chapter. The fourth chapter addresses the core of the research problem i.e. estimation of the recreation value of the lake. The fifth chapter is the summary of the findings.

2

Economics of Tourism

I

INTRODUCTION

The chapter is an attempt to blend economics of tourism and valuation of outdoor recreation part I of the chapter focuses on meaning and concept of tourism while economic methods for valuing outdoor recreation constitutes the content of the part II.

Tourism—Concept and Meaning

Tourism is a leisure activity. It exists with its opposite, i.e. regulated and organised work. It involves the movement of people to and their stay, at different places. It has three important aspects—destination, accommodation and transportation. It is concerned with spatial conditions lake the location of tourist areas and the mobility of people between place to place. Tourism is closely related to the structure, form, use and conservation of landscape and related socio-economic consequences. Man has been an untired traveller from time immemorial. By the way the motive of travel have undergone changes from time to time. In early times, the motivations for travel were basically three: conquest, trade and pilgrimage. But now-a-days it is concerned with pleasure, holiday, travel and going or arriving somewhere. Tourism, as a significant social phenomenon, involves a temporary break with normal routine to engage with experiences that contrast with everybody the explorer, the pilgrim, the monk, the merchant, the student, the missionary, the hermit, the refugee, the conqueror, the cure seeker etc., life with the mundane can be cited as prototypes of the modern tourist.

According to Gilvie (1933) a tourists is a person who satisfies two conditions—first, he is away from home for any period less than a year and second, while he is away, he spend in the place he visit, without earning there. Tourism embraces all movement of people outside their community for all purposes except migration or regular daily work. The reasons can be attending conferences and business purpose (Licorisch 1953).

Cohon defines that "A tourist is a voluntary temporary traveller. He is travelling in expectation of pleasure from the novelty and change experienced in a relatively long and non-recurrent round trip". Bukart and Medlik (1976) "Tourism denotes the temporary and short-term movement of people to destinations outside the places where they normally live and work, and their activities at those destination.

Hunzikar and Krapf (1951) state tourism "as the sum of the phenomena and relationships arising from the travel and stay of non-residents, in so far as it does not lead to permanent residency and any earning activity".

Although a number of definitions encompassing diversified activities concerning a complex phenomena like tourism have been coined but till now a uniformly acceptable definition is yet to be formulated.

Economic Contribution of Tourism

Tourism is one of the disguised blessings attributes to most developing nations. It can act as the pivot of vehicle for economic development. It has been recognised as one of the biggest component of the tertiary sector. It constitutes a major item in the world trade, growing at a faster rate than the world trade in tangible goods. It has become a highly organised industry representing 11 per cent of World's Gross Product. World's revenue from tourism has increased from $70 billion in 1960 to $1500 billion in 1995.

The stream of economic benefits flowing from tourism are discussed below:

The Multiplier Effect of Tourism

One of the important economic features of tourism is that income earned in places of 'residence' is spent in places 'visited'. Tourism, thus, is instrumental in transferring vast sums of money from the 'income generation' countries to 'income receiving' countries. The money spent by tourists tends to percolate through many levels. Tourist expenditure stimulates domestic flow of rupee income through several streams. It generates additional income at each round of spending. There is, thus a multiplier effect. The word 'multiplier' expresses the numerical co-efficient showing the actual change in income brought about by some change in input or investment. It is the number by which a change in input must be multiplied in order to determine the resulting change, in income not excluding the turnover of initial expenditure.

The formula is

$$K = \frac{1}{1 - C/Y} \quad (1)$$

where K stands for the multipliers, C stands for consumption and Y for income.

The multiplier K is dependent on the relationship between a change in consumption C and a change in income Y. Let us illustrate this. Assume that the tourist expenditure in national is Rs. 5 lakhs particular period of time, the change in income (Y) will be Rs. 15 lakhs and the change in consumption (C) will be Rs. 15 lakhs. Putting the values in equation 5, K becomes

$$\frac{1}{1 - 10/15} = 3.$$

Tourist expenditure and associated investment results in significant induced income generation in the host economy in addition to its direct and indirect effects on the tourism sector and its input suppliers. The multiplier effects of tourist spending constitute one of the better researched areas within the tourism economies literature (Archer 1977, Sheldon 1990, Fletcher and Archer 1991). Two main approaches have been used to estimate tourism multiplier values: (i) the Keynesian multiplier model and

(ii) the input-output technique. Both approaches can be used to estimate the multiplier valued associated or government spending. The Keynesian model requires less data but provides less information about the interrelationships within the economy than the input-output approach.

A basic equation which can be used to estimate multiplier values using the Keynesian model is given as:

$$k = \frac{1-l}{1-c(1-t_i)(1-t_d-u)+m} \quad (2)$$

where l are first round leakages, c is the marginal propensity to consume, t_j is the marginal rate of indirect tax, t_d is the marginal rate of direct tax, u is the marginal rate of transfer payments and m is the marginal propensity to import. The formula has been modified from time to time to measure increases in investment. The first-round leakages can be tailored to the specific nature of the initial injection and the definition of income to be measured (Sinclair and Sutcliffe, 1978, 1988). The incorporation of first-round propensities applied at the disaggregated firm level (Archer and Owen, 1971) and propensity values which are appropriate for different time periods, facilitate the estimation of short and long-run multiplier values (Sinclair and Sutcliffe, 1989). Firm-specific propensities were used in Milne's (1987) study of differential multiplier values for tourist accommodation and handicraft firms and tour operators in the Cook Islands. The model is:

$$Y_a = \frac{W(1-h-t_w)+P(1-t_p)+F(1-t_w)\sum_{j=1}^{l} S_{ai}Y_t}{D_a} \quad (3)$$

where Y_a is the local income creation coefficient by establishment, W is gross wages and salaries of residents, h is national insurance and other deductions form wages and salaries, t_w is direct taxes, P is local profits, t_p is taxes on profits, F is rent paid to residents. S_{ai} is business purchases from the ith business, Y_I is destination income generation coefficient for the ith business and D_a is total turnover of the business. The results showed that smaller firms were associated with higher multiplier values than larger establishments owing largely to their relatively low propensities to import and to repatriate profits overseas.

The input-output technique requires more detailed information about inter-sectoral transactions in order to estimates the impact of tourist spending on different sectors within the economy. Using the matrix of technical coefficients for the economy. A and the vector of gross output, X, the vector of final demand Y,

The equation is,

$$Y = X (I-A) \quad (4)$$

Where I is the identify matrix. Equation (4) can be rearranged to give

$$X = (I-A)^{-1} Y \quad (5)$$

So that $\Delta X = (I-A)^{-1} \Delta Y$ (6)

This basic model equation (6) can be used to obtain different categories of multiplier. Type I income multipliers are the ratio of direct and indirect income generated by the initial injection to the direct income generated. Type II multipliers are the ratio of total direct, indirect and induced income generated to the direct income these are obtained by considering household income and expenditure as endogenous within the transactions matrix. Type III multipliers are similar to type I and type II multipliers. But these multiplier take account of alterations in consumption as income increases (Fletcher and Archer, 1991). The multipliers can be estimated calculating the total income generated within each sector of the economy per unit increase in the exogenous final demand for the output of the sector.

The input-output technique assumes proportionality in inter-sectoral relationships and constant returns to scale but has been refined to incorporate the effects of the changes in consumption patterns that occur as income rises, (Sadler et al. 1973). The model is

$$(I-A)^{-1}\left[I-\sum_{h=n-l}^{p} C_{h(I-A)B_h(A)^{-1}}\right]^{-1} Y = X \qquad (7)$$

where X is a vector of output by sector, C_h is one of p matrices giving household consumption patterns, B_h is one of p diagonal matrices giving household earning categories and Y is the vector of final demand.

Wanhill (1988) in his model of tourist expenditure in Mauritius has introduced capacity constraints. The equation is

$$\Delta X = (I-RA)^{-1} R\Delta T \qquad (8)$$

where R is a matrix of capacity constraints ranging from 0 to 1 and ΔT is a matrix of the change in tourist spending. If labour market constraints are also incorporated into the model, the change in employment generated by the change in tourist spending, ΔL, can be written as

$$\Delta L = NE^{*} (I-RA)^{-1} R\Delta T \qquad (9)$$

where E* is a partitioned matrix of employment coefficients and N is an employment restrictions matrix which takes account of job creation in the destination area. The application of the model demonstrated the importance of allowing for capacity constraints as the calculated income multiplier values were up to 28 per cent lower than those resulting from the unconstrained model and the employment multiplier values were up to 34 per cent lower (Wanhill 1988, Fletcher and Archer 1991).

Contribution of Tourism to National Income

Tourism has significant contribution to the national income of a host country. Tourists have to pay for different types of services and goods in the host country. On a global scale, tourism receipts have been estimated to be more than one per cent of the GNP. In case of India, it has been estimated that tourism earnings account for 2.2 per cent of the net national income. In countries like Spain, Greece and Switzerland, tourism receipts account for 3 to 7 per cent of their GNP. In case of Maldives, tourism accounts for 74 per cent of the GNP.

Harbinger of Employment Opportunities

Tourism is a service industry and occupies a predominant place in providing employment opportunities to thousands of people. Tourism provides direct and indirect employment both in the skilled and unskilled categories. A large number of economic activities come under tourism. These include the hospitality industry, air transport, surface transport, travel agencies, tour operators, guide services, souvenir establishments agencies providing safety and security, etc. Besides there are good number of supporting activities like supply of goods to the hotels, to the airlines, and to the railways, etc. In India tourism provides a vast spectrum of jobs ranging from highly trained managers of 5 star hotels to room boys, sales girls, handicraft artisans and transport works. With fast growth of tourism, new horizons job opportunities open up for unemployed or partially employed young men and women. The airlines, travel agencies and tour operators need thousands of young people with varied skill. There is also enough scope for self employment in a variety of ancillaries.

Tourism generates employment in the formal sector but also in informal sector (Elkan 1975). This has been cited as one of its key advantages for developing countries (de Kadt 1979). Empirical studies have confirmed that the level of employment in tourism activities is high. These account for 0.5 million jobs in Spain (Sinclair and Bote Gomez, 1996) and around five million in India (ESCAP, 1991). The direct employment in tourism as a share of total employment, in Spain was around 6.2 per cent at the beginning of the 1990s (Bote Gomez 1993), compared with 2.1 per cent of formal sector jobs in India, 0.9 per cent for Sri Lanka (Attanayake et al. 1983), 1.3 per cent for Zimbabwe (EXA International, 1993). It is important to highlight the gender disparity regarding job opportunities.

Gender Disparity in Tourism

It is important to highlight the gender disparity in regarding job opportunities. Most of the top jobs are undertaken by men while the lower paid, part-time and seasonal jobs are predominantly filled by women (Sinclair, 1997). However, it is interesting to note that the growth of tourism has created new opportunities for women. In case of Cyprus, Cypriot women have usually carried out cleaning, bed-making and cooking in small guesthouses while men have been responsible for financial and social interactions with tourists. Women from Eastern Europe have worked as croupiers in casinos (Scott 1997). Large hotels which catering to foreign tourists generally provide employment opportunities to young and relatively well educated Cypriot women. Women in Barbados (Levy and Lerch 1991) and Sri Lanka (Samarasuriya, 1982) have been employed in tourist guesthouses. In Bali, women have established small-scale, tourism-related businesses such as home stays or souvenir shops, although it is generally unacceptable for married women employed in tourism activities. (Cukier et al., 1996; Long and Kindon, 1997). The growth of international tourism has also been accompanied by considerable increases in employment in prostitution in developing countries. (Lee, 1991; Chant, 1997).

Effects of International Tourism

Tourism has been increasing at a faster rate than world trade. The percentage of tourism earnings in the total exports of many countries has been increasing year after year. It has been universally accepted that as soon as a country is able to earn foreign exchange from tourism which is a minimum of 10 per cent of the merchandise exports, that country can be called a 'Tourism' country. Among the three most important components of Trade in services, namely, transportation, tourism and insurance, the net foreign exchange gains have always been positive in case of tourism, while in the case of the other two components, the net gain has mostly been negative.

Tourism appeal to developing countries is based, in large part, on its provision of foreign currency earnings and corresponding alleviation of the balance of payments constraint. (Thirlwall, 1979; Thirlwall and Mureldin-Hussein, 1982). International tourism has been one of the most important sources of foreign currency, in terms of both absolute value and growth, in many developing countries (UNCTAD, 1973; English, 1986; Lee, 1987; Lea, 1988; Euromonitor, 1997). In Kenya, tourism has overtaken the traditional primary commodity sectors of coffee and tea to become the country's main foreign currency earner (Sinclair, 1991b). It is the second most important 'export' in The Zambia (Dieke, 1993) and, in Egypt, foreign currency earnings has exceeded by remittances from abroad (Huband, 1997). Asian countries such as Thailand and Indonesia are renowned for their high levels of tourism receipt and many small-island economies, like Fiji (Varley; 1978), Jamaica (Curry, 1992). Bermuda (Mudambi, 1994; Archer, 1995), the Maldives (Sathiendrakumar and Tisdell, 1989) and the Seychells (Archer and Fletcher, 1996), depend upon tourism activities (Conlin and Baum, 1995).

It is clear that much closer links between the tourism sector and the local economy would raise the farmer's income and employment generating capacity and in this context, increased purchasing form local producers of the agricultural and marine products which tourists consume would be particularly useful.

There are a large number of socio-economic benefits from tourism which can't be quantified. It provides, stimulus to the development of local handicrafts generation of exports, development of local cultural activities, there is development of tourism related linkage, ancillaries and self-employment activities. Further tourism provides a major thread to knit this vast nation to a cohesive feeling of togetherness by overcoming the barriers of caste, creed, religion, language and distance.

Through tourism people can better understand and appreciate the culture, history, geography, social, educational, political and economic systems of other countries. This realisation will certainly lead to greater understanding between and among the people of various countries and in maintaining world peace. There are a number of occasions when there is direct cultural and social contact between and among the tourists and the citizens of the host country.

There has to be an element of human relationship in all tourism activities. The human aspect of tourism has considerably improved the quality of life of people. As mentioned earlier besides its immense economic benefits, tourism has been health oriented and education oriented. Tourism helps in understanding the viewpoints of others. People started realising that they have to live with differences. That is in fact, the basis of toleration and coexistence. Tourists can imbibe these values through tourism. He can become a better human being. He can realise the good values of his own country through comparison with other countries. Better citizens of the world, thus are produced through tourism. Tourism has been responsible for the preservation of some of the rare flora and fauna. Through systematic tourism planning many countries of the world have taken special measures to preserve and protect the flora and fauna. There are some negative aspects of tourism. These are, however, marginal and relate primarily to overcrowding of a tourist spot, natural or man-made. However, even these marginal negative factors must be overcome by a systematic tourism promotion policy.

Tourism Earnings Instability and Economic Growth

Tourism sometimes is perceived as a high-risk option for developing countries. The earnings may not be stable always. For example, relatively high elasticities of demand with respect to changes in inflation, exchange rates or political instability are often associated with considerable changes in earnings. In Kenya, real dollar earnings from tourism, both total and per tourist, were higher during the 1970s than during 1980s, owing partly to the fact that contracts between tour operators and hoteliers were often negotiated in terms of the depreciating Kenyan shilling (Sinclair, 1990, 1992). Instability brings various adverse effects in the host economy (Rao, 1986).

An alternative set of theories, based on Friedman's (1957) permanent income hypothesis of consumption, proposes the contrary view that export earnings instability is beneficial to growth. It is argued that unstable export earnings are likely to be perceived as transitory rather than permanent income, bringing about increases in savings, rises in investment and higher growth (Knudsen and Parnes 1975). The implication is that instability in tourism earnings is not a problem.

There is little evidence on the scale or effects of instability in tourism earnings. However, research on a sample of developing and newly industrialising countries with relatively large tourism sectors was undertaken by Sinclair and Tsegaye (1990) using the following definitions of instability:

$$l_1 = 100/n\sum_{t=1}^{n}|(X_t - X_t^*)|/X_t^* \quad (1)$$

and $$l_2 = 100\left[1/n\sum_{t=1}^{n}\{(X_t - X_t^*)/X_t^*\}^2\right]^{1/2} \quad (2)$$

where X_t is the actual value of export earnings in time period t, X_t^* is the trend value of earnings from a regression of earnings against time and n is the number of observations. Additional instability measures were calculated on the basis of deviations from a five year moving log average. Positive correlation coefficients between the instability measures for tourism and merchandise export earnings for the period 1960 to 1985 indicated that tourism had a net destabilising effect on the earnings of Fiji, Jamaica. Cyprus, Greece and Turkey Coefficients which were insignificant or positively signed for one measure but negatively signed for another indicated an insignificant or ambiguous effect for India. Mexico, Morocco, Singapore, Thailand and Tunisia, Buckley's (1993) study of the later 1978-89 period found a net stabilising effect for Cyprus, Greece and Turkey. Further, research indicating the level and trends in dependence of specific developing countries on particular tourist origin countries and on the associated degree of earnings instability, would be useful.

The possibility of using portfolio analysis to determine the optimal mix of tourism and other exports, or of different nationalities or types of tourist, has been suggested by Board et al., (1987) and Buckley (1993). Portfolio analysis is based on Markowitz's (1959) model for risk reduction in financial asset holdings and its application to tourism and other exports involves treating export earnings analogously to the returns on assets and the instability of earnings as a measure of risk. The model can be used to assist countries to select their desired trade-off between the level of earnings and their degree of instability. Thus, developing countries which are prone to adverse effects of instability may aim to attain a portfolio of tourism and other exports which involves a lower risk-return trade-off than industrialised countries whose larger reserves of foreign currency enable them to cope more effectively with any problems arising from instability.

Cost of Tourism

Tourism is perhaps the biggest global business activity after oil. Both the natural environment in the form of land, water, plants and animals, and the man-made environment constitute the attraction, which tourists look for in a destination.

Tourism comprises of the complete system of nature – the universe, the space and the galaxy which includes the man and his activities, wildlife, mountains and valleys, rivers and waters, forests and trees social and cultural system, flora and fauna, weather and climate, sun and the sea. The whole system requires an environmental and ecological preservation which can be expressed in terms of an equation.

$$(N+W+M)^{EE} = \text{Tourism} \quad (3)$$

Where, N = Nature, W = Wild life, M = Man and this activities, E = Environment, E = Ecology

This equation can be further simplified as Nature (wild life + man)^{EE} = Tourism (Nagi, 1990). These are damages to the natural environment notably by the construction and expansion of hotels, shops etc. and the contamination of rivers and beaches (Hollaway, 1994). An area of scenic beauty attracts greater number of tourist, so more and more of the natural landscape is lost to tourism development ultimately resulting in permanent damage to the landscape and site. This is the stage called 'senity' in the life cycle of a tourist spot. In 1979, the OECD, highlighted four main categories of environmental stress created by tourism. These are:

Permanent environmental restructuring (major construction works such as highways, airport and resorts), waste product generation/biological and non-biological waste which can damage fish production, create health hazards and reduces the attractiveness of a tourist destination, direct environmental stress caused by tourist activities (destruction of coral, reefs, vegetation, dunes etc. by the presence and activities of tourists) and effects on the population dynamics (migration increased urban densities accompanied by declining populations in other rural areas).

Increase tourism activities will have both direct and indirect negative impacts on natural environment. Negative environment impacts include water, air and noise pollution, damage to wildlife and vegetation etc. Use of motor boats on inland waterway adversely affects the natural flora and fauna.

Increasing international tourist inflow at a place leads to increasing population densities and greater transportation activities, which cause pollution of the environment. The Natural Resources Defence Council, a non-profit environmental research organisation reported that 2008 public beaches in 14 coastal states were closed in 1996 due to contaminated water. High bacteria levels from human and animal wastes were the main reasons for most of the closings. Some of the more specific negative environmental impacts of tourism are the following:

Hunting and fishing damage the wildlife environment, sand dunes are eroded by excessive use, vegetation can be destroyed by walkers, campfire can destroy forests, the construction of tourism superstructure can be at the cost of the aesthetics of the land, and the improper waste disposal can damage the environment.

The environmental damage is aggravated by frequency of visitors. The flora and fauna gen no ample change for regeneration and the ecological balance is likely to be disturbed. Water supply became insufficient and polluted. Different kinds of tourism activities affect the natural and build environment. The ecological and environmental impact of tourism has been discussed below.

Tourism affects the water environment in two ways: Garbage and sewage is released into lakes, rivers and beaches thereby polluting it. The marine water bodies are polluted by release of oil from cruise ships and ferry boats etc. In the fresh water bodies contamination of waters leads to health hazards and ultimately to destruction of aquatic plant and animal life. On the other hand the increased toxicity due to oil spell leads to contamination of sea food and affects the marine food chain.

Due to increase tourism operation the tourists inflow to destination increases. Increased use of vehicles like motorcar, ship, train, aeroplane etc. gives rise to air and noise pollution. Air pollution affects plant life and noise pollution affects the aesthetic beauty of the place.

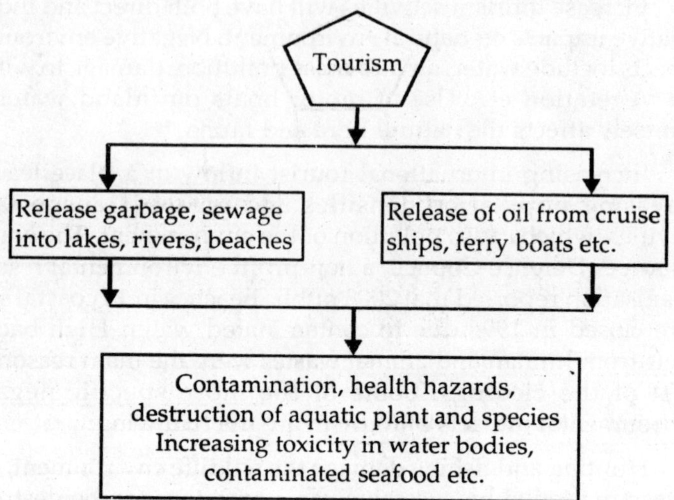

Fig. 2.1: Effect of Tourism on Water bodies

Source: **Anita, 1997.**

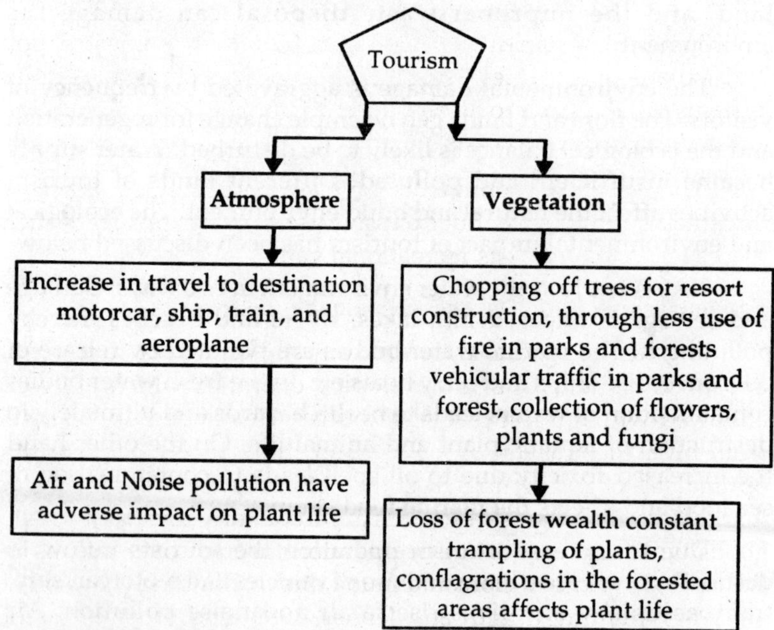

Fig. 2.2: Effects of tourism on atmosphere and vegetation

Source: **Anita, 1997.**

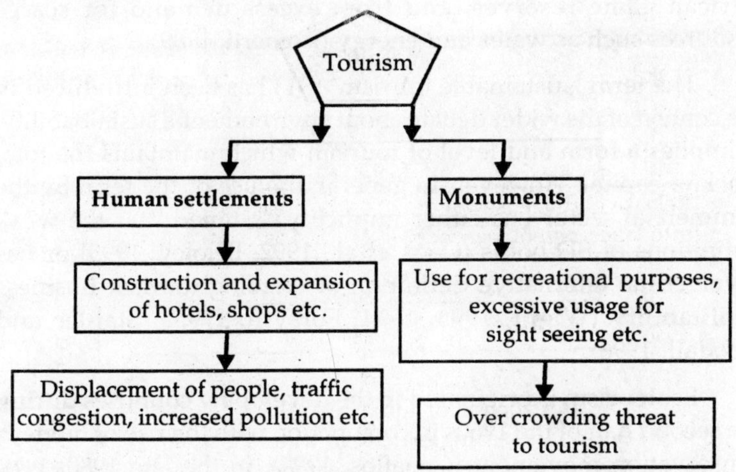

Fig. 2.3: Effects of Tourism on Human Settlements and Monuments

Source: **Anita, 1997.**

With enhancement of tourist operation there is construction and expansion of hotels, shops, roads and other public utility systems. All these led to displacement of people, traffic congestion and increased pollution etc.

The historical monuments have always remained as places of attraction for tourists. But excessive use or overexposure of such places leads to loss of its recreational value.

Eco-tourism: A Path Towards Sustainable Tourism

Increasing specialisation in tourism would, at first sight, appear to be beneficial to developing countries which are well endowed with environmental resources 'demanded' by tourists from high income nations. However, they rarely achieve the optimal use of and returns from their assets owing to problems of market failure. Natural resources are frequently unpriced public goods which are subject to degradation through over-use, as in the case of the loss of flora and fauna due to climbing (Pawson et al., 1984), hunting (Smith and Jenner, 1989) and other forms of tourist activity (Andronikou, 1987). Problems also arise from off-road use of vehicles (Sindiyo and Pertet, 1984) as in the case of

African game reserves, and from excess demand for scarce resources such as water and energy (Romeril, 1989).

The term 'sustainable tourism' (ST) has been introduced in the context of the wider debate about environmental sustainability. It implies a form and level of tourism which maintains the total stock of capital. However, in general, the use of the term by the commercial sector has either implicitly assumed that the weak definitions of SD holds (Cook et al., 1992; Beioley, 1995) or has involved an alternative definition of ongoing tourism business profitability (Wight, 1993, 1994; Forsyth. 1995a; Stabler and Goodall, 1996).

Ecotourism is originated in the developing countries during the second half of the 1960s in conjunction with the rise of modern conservation movement (Ceballos, 1993). In the late 1980s ecotourism was an unknown entity that was just beginning to emerge in the popular lexicon. Its growth was spurred by the ongoing debate over tourism and the environment and as a direct result of the enthusiasm for ecological sustainable development (Ecological Sustainable Development Working Groups. 1991). The first example of ecotourism business operation was noticed in the Peruvian Amazan in 1979, where an operator was in the process of adopting Kenya's tree tops hotel design to open up rain forests to tourists. Once the visitors discovered the delights of the rain forests by becoming an observant and sensitive. By the early 1990s the term ecotourism term had become a positioning statement and politically correct from mass tourism.

Ecotourism is defined as the activities of persons travelling to and staying in places outside their usual place of residence for not more than one consecutive year for leisure, business and other purposes constitute tourism. Such visits should intend to enjoy nature in a most environment friendly manner without any adverse impact on the ecosystem.

It embraces "environmental-friendly", "community-friendly", and "market-friendly" tourism. Eco-tourism refers to niche market for environmentally aware tourists interested in observing nature. Tourism is an environment dependent industry. Eco-tourism is the largest expression of this relationship. Tourism has benefited the environment by stimulating measures to protect physical features of the environment, historic sites, same monuments and wildlife. Recreation and tourism are normally the primary objective of establishing and developing national parks and many other types of protected areas. These natural areas become major attractions and constitute the basis for what is known as "Nature-Tourism" or "Eco-Tourism".

Eco-tourism is an offshoot of the wave of environmental awareness. The main idea is to make it as a tool for the protection of natural ecosystems by giving them a socio-economic value in their original state. One value of Eco-tourism is its potential for promoting alliance among environmental, conservation and development interest. It recommends great public participation in planning and decision-making concerning living resources use. It proposes environmental education programmes and campaigns to build support for conservation (Birundha, 2003).

The global importance of eco-tourism, its benefits as well as its impact was recognized with the launching of the year of eco-tourism (IYE) by the United Nations General Assembly. Recently it has been regarded as the panacea that enables us to aggressively seek tourism dollars with no obvious damage to ecosystem.

Eco-tourism is responsible travel to natural areas that conserves the environment and sustains the well being of local people. According to World Bank Conservation Union (IUCN, 1996) eco-tourism is "Environmentally responsible to travel to natural areas, in order to enjoy and appreciate nature, and accompanying cultural features (both past and present) that promote conservation, have a low visitor impact and provide for beneficially active socio-economic involvement of local peoples". Conservation, sustainability and biological diversity are the three inter-related aspects of eco-tourism.

It can be stated that it is a type of tourism that is sustainable.

- Is integrated into surroundings.
- Is environmentally sound, and no degradation of the resource.

- Has developed in natural environment.
- Has a characteristic demand, motivated by the natural environment.

Eco-tourism can take many forms and magnitudes. For example, losing one self in a beautiful natural forest or landscapes – watching animal, birds and trees in a forest, corals and marine life in a sea engaging in trekking, boating or rafting.

To sum up, it can be said that tourism is essential. But tourism, economic development and environment have to go hand in hand and must develop a symbolic relationship. Thus, it is suggestive to develop sustainable tourism from a regional, sub-regional or even from an international standpoint.

STATE OF TOURISM IN ORISSA

Orissa comprises of 4.74 per cent of India's landmass and 36.80 million people (2001 census), accounts for 3.58 per cent of the population of the country: Nearly 85 per cent of its population live in the rural areas and depend mostly on agriculture for their livelihood. The state has rich forestry, natural land scopes, abundant fishery resources, mineral resources and plentiful resources. The state has wonderful base for tourism. The Sun Temple at Konark, the Lingraj Temple, the Jagannath Temple at Puri, the Golden beach of Puri and Gopalpur, the rich biodiversity and avifauna of lake Chilika enriches the tourism base of the state. This section is a brief discussion on this.

Inflow of Tourists to Orissa

The state Department of Tourism has worked out the data relating to the tourist arrivals. Table 2.1 presents the inflow of tourists into Orissa: during 1992-2004. it reveals that the total inflow of tourists was 24,97,985 in 1992, which increases to 41,54,353 in 2004. If we look into the annual percentage change we find that it has fluctuated over the period. Tourist's inflow into Orissa has increased except in the year 1999.

Table 2.1: Inflow of tourists in Orissa during 1992-2004

Year	*Domestic*	*% of change*	*Foreign*	*% of change*	*Total*	*% of change*	*Share of domestic tourist on compared to the total tourist (%)*	*Share of foreign tourist on compared to the total tourist (%)*
1992	24,71,343	–	26,639	–	24,97,985	–	98.93	1.07
1993	25,38,875	2.7	24,856	6.7	26,63,731	2.6	99.03	0.97
1994	25,94,992	2.2	26,024	4.7	26,21,016	2.2	99.01	0.99
1995	26,97,365	3.9	28,201	804	27,25, 566	4.0	98.94	1.06
1996	27,73,245	2.8	34,303	21.6	28,07,548	3.0	98.77	1.23
1997	28,28,131	2.0	35,081	2.3	28,63,212	2.0	98.77	1.23
1998	28,61,788	1.2	33,101	-5.6	28,94,889	1.1	98.86	1.14
1999	26,91,840	-5.9	25,758	-22.2	24,17,598	-6.1	99.06	0.94
2000	28,88,392	7.3	23,723	-7.9	29,12,115	7.2	99.18	0.82
2001	31,00,316	7.3	22,854	-3.9	31,23,170	7.2	99.20	0.74
2002	34,13,352	10.1	23,034	0.8	34,36,386	10.3	99.33	0.67
2003	37,01,250	8.4	25,020	8.6	37,26,270	8.4	99.33	0.67
2004	41,25,536	11.5	28,817	15.2	41,54,353	11.5	99.31	0.69

Source: Statistical bulletin 2004, Department of Tourism and Culture, Government of Orissa.

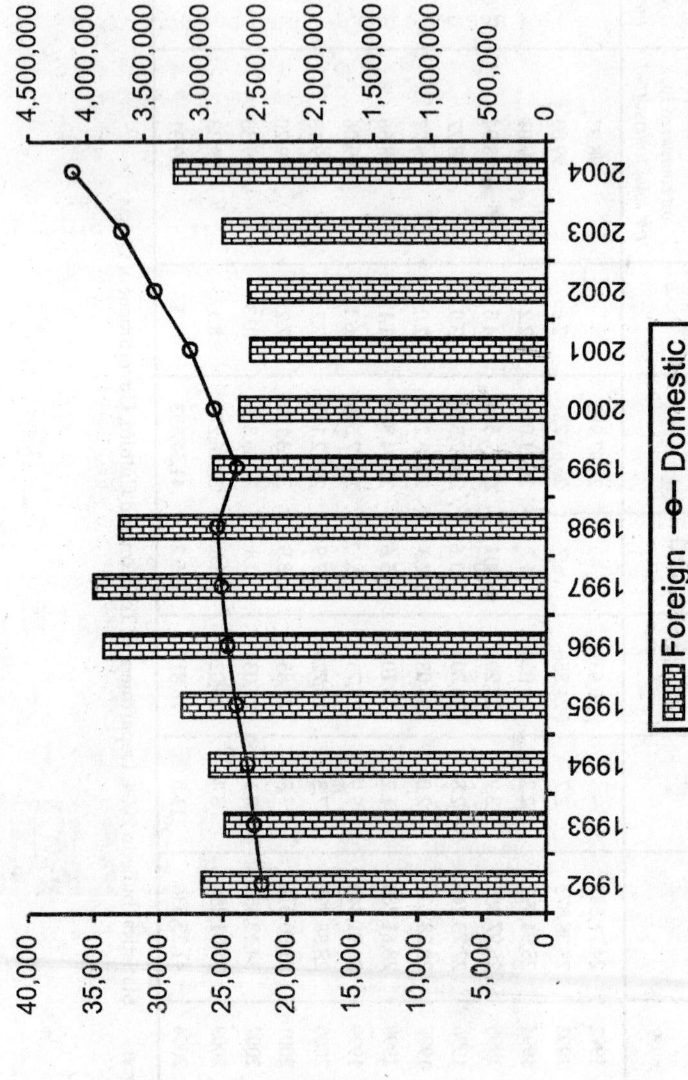

Note: **In this figure left side scale shows Foreign Tourists & right side scale shows Domestic Tourists.**

Fig. 2.4: Number of Domestic and Foreign Tourists arrival to Orissa from 1992-2004.

Due to the Super Cyclone in 1999 there was a decline in the tourists inflow into Orissa. Inflow of domestic and foreign tourists into Orissa in 1999 have declined by 5.9 and 22.2 respectively when domestic tourists inflow increased in 2000 by 7.3 per cent at that time foreign tourist inflow agencies has declined by 7.9 per cent.

From the Table 2.2 *(See on next page)* it can inferred that domestic tourists inflow into Orissa has increased in a continuous manner from 1992 but there was a wide gap in 1999. But in the case of foreign tourists inflow into Orissa there was an increased from 1994 to 1996 and then again a declining trend in 1997 to 2001, but registering an increased of 21.6 per cent for the year 1996. The inflow of tourists, however shown an erratic and zigzag trend during 1992 to 2003 but a steady increase during 1994-1997, and a steady fall during 1998-2001. The percentage growth in 1996 was 21.6 per cent as against the previous year. But after 1996, it reveals a downward trend until it reaches 7.9 per cent and 3.7 per cent in 2000 and 2001 respectively.

Table 2.2 shows that in the month of December tourist inflow into Orissa is high. It also reveals that the largest inflow of domestic and foreign tourists is highest during December. In December the percentage of domestic tourist registered 14.01 per cent, whereas foreign tourist marked 16.36 per cent of the total tourists inflow during the year. It can be said that for domestic tourists, peak season is October to January whereas for the foreign tourist the peak season is November to March.

From Table 2.3 *(See on next page)* one can get the idea of the share of Orissa in India's foreign tourism. Among the total foreign tourists visiting India, only 0.86 percent came to Orissa in 2004. It is seen that per cent of foreign tourists inflow into Orissa have been declining since 1996. In 1996 the per cent of foreign tourists visiting Orissa was 1.50 and in 2004, it declined to 0.86 per cent. It can be inferred that Orissa's share of foreign tourist in India is disheartening (i.e. less than one per cent).

Table 2.2: Month-wise tourists inflow into Orissa 2003

Month	*Number of tourist*			*Proportion of total tourist (in %)*	
	Domestic	*Foreign*	*Total*	*Domestic*	*Foreign*
January	3,00,204	3,546	3,03,750	8.11	14.17
February	2,57,347	3,203	2,60,550	6.95	12.50
March	2,63,767	2,857	2,66,624	7.13	11.42
April	2,52,817	1,586	2,54,403	6.83	6.34
May	2,77,075	918	2,77,993	7.49	3.67
June	2,77,119	878	2,78,017	7.49	3.59
July	3,97,343	1,086	3,98,429	10.74	4.34
August	2,71,781	1,364	2,73,145	7.34	5.45
September	2,52,421	980	2,57,401	6.81	3.92
October	3,34,643	1,516	3,36,159	9.04	6.06
November	2,98,270	2,973	3,10,243	8.06	11.88
December	5,18,463	4,093	5,22,556	14.01	16.36
Total	**37,01,250**	**25,020**	**37,26,270**	**100.00**	**100.00**

Source: Statistical Bulletin, Government of Orissa, Department of Tourism and Culture, 2003.

Table 2.3.: Foreign Tourists arrival in Orissa and India (1992-2003)

Year	*Number of Foreign Tourists arrival to India*	*Number of Foreign Tourists arrival to Orissa*	*Share of foreign Tourists arrival to Orissa (in %)*
1992	18,67,651	26,236	1.42
1993	17,64,830	24,856	1.40
1994	18,86,433	26,024	1.38
1995	21,23,683	28,201	1.32
1996	22,87,860	34,303	1.50
1997	23,74,094	35,081	1.48
1998	23,58,629	33,101	1.40
1999	24,81,928	25,758	1.04
2000	26,49,378	23,723	0.90
2001	25,37,282	22,854	0.90
2002	23,61,587	23,034	0.98
2003	27,50,290	25,020	0.91
2004	33,67,980	28,817	0.86

Source: Statistical Bulletin 2003, Department of Tourism and Culture, Government of Orissa.

Visit of Domestic Tourists

Table 2.4 *(See on next page)* reveals that the share of domestic tourist inflow into Orissa was 21.48 per cent from West Bengal, 4.42 per cent from Andhra Pradesh, 2.82 per cent Jharkhand 2.62 per cent from Chhatisgarh, 2.79 per cent from Maharastra, 2.27 per cent from Uttar Pradesh and 2.12 per cent from Delhi in 2003.

Visit of Foreign Tourists

Table 2.5 *(See on page 43)* shows that U.K, U.S.A, France and Germany are the major tourist generating markets for Orissa during the year 2003, with 3,079 tourists (12.30%), 2215 tourists (8.85%), 2,562 tourists (10.24%) and 2,186 tourists (8.74%) respectively out of the total tourist visit of 25,020 followed by Japan (7.62%), Italy (7.21%), Netherlands (5.75%), Australia (3.95%), Switzerland (3.04%) and Korea (2.57%).

Table 2.4: Domestic Tourists in 2003

Name of the state/UT	*No. of Tourist during 2003*	*Proportion to total (in %)*
West Bengal	7,95,050	21.48
Andhra Pradesh	1,63,644	4.42
Bihar	73,283	1.98
Jharkhand	1,04,356	2.82
Madhya Pradesh	56,820	1.54
Chhatisgarh	97,008	2.62
Maharastra	1,03,115	2.79
Uttar Pradesh	84,143	2.27
Uttaranchal	27,661	0.75
Tamilnadu	58,094	1.57
Karnatak	33,426	0.90
Gujarat	31,712	0.86
Rajasthan	19,244	0.52
Assam	18,319	0.49
Punjab	9,436	0.25
Haryana	7,242	0.20
Kerala	7,302	0.20
New Delhi	78,546	2.12
Orissa	18,76,571	50.70
Other states/UT	56,258	1.52

Source: Statistical Bulletin 2003, Department of Tourism and Culture, Government of Orissa.

Table 2.5: Visit of Foreign Tourist to Orissa

Sl. No.	*Name of the Country and Region*	*Number of tourist (2001)*	*Number of tourist (2002)*	*Number of tourist (2003)*
1	*2*	*3*	*4*	*5*
North America				
1.	U.S.A.	2298	2012	2215
2.	Other	558	838	804
	Total	**2856**	**2850**	**3019**
Central and South America				
	Total	259	236	200
Australia				
1.	Australia	967	929	990
2.	Other	156	290	235
	Total	**1123**	**1219**	**1225**
Western Europe				
1.	France	1804	2019	2562
2.	Germany	2022	1977	2186
3.	Italy	1553	1360	1805
4.	Netherlands	1046	1195	1439
5.	U.K.	3367	3012	3079
6.	Other	3381	3765	3557
	Total	**13173**	**13328**	**14751**
Eastern Europe				
	Total	653	741	629
West Asia				
	Total	3366	372	353
South Asia				
1.	Bangladesh	393	449	354
2.	Nepal	213	371	450
3.	Other	204	228	195
	Total	**810**	**1048**	**999**

(Contd...)

1	2	3	4	5
South East Asia				
	Total	421	498	716
East Asia				
	Total	2977	2518	2874
Africa				
	Total	**246**	**224**	**254**
	Grand Total	**22854**	**23034**	**25020**

Source: Statistical Bulletin, Department of Tourism and Culture, 2003, Government of India.

Earnings from Tourism

Orissa Government earned foreign currency worth of 34.18 crore from foreign tourists and Rs. 1381.56 crore from domestic tourists in the year 2003. The total earning from tourism in 2004 was Rs. 1415.74 crore (Table 2.6).

Table 2.6: Inflow of Money through Tourists expenditure in Orissa during 2004

Types of Tourist	*No. of Tourist*	*Average duration of stay (in days)*	*Per capita expenditure per day in (Rs.)*	*Estimated inflow of money through Tourist expenditure (Rs. in crore)*
Domestic	41,25,536	5.2	644.00	1381.56
Foreign	28,817	10.6	1119.00	34.18
Total	**41,54353**			**1415.74**

Source: Statistical Bulletin, Department of Tourism and Culture, 2004, Government of India.

The average expenditure incurred by a foreign tourist in Orissa in 2004 Rs. 1119 excluding the expenditure in reaching or leaving the state. The average duration of stay of a foreign tourist in the state was 10.6 days. The average per capita daily expenditure of domestic tourist in the state was Rs. 644.00 in 2004 excluding the expenditure on reaching or leaving the state. The average duration of stay of a domestic tourist in the state was 5.2 days (Table 2.6).

It is found that the continuous increase in foreign currency in Orissa through foreign tourists have got a break in 1999 due to fall in tourist inflow due to the Super cyclone. However presently it is increasing (Table 2.7).

Table 2.7: Foreign Exchange earnings in India/inflow of money through tourist expenditure in Orissa since 1990

Year	*Estimated travel receipts in India (Rs. in crore)*	*Inflow of Money through tourist expenditure in Orissa (Rs. in crore)*
1990	2612.50	5.98
1991	4892.00	6.33
1992	6060.00	6.40
1993	6509.00	6.43
1994	7103.50	7.44
1995	8640.00	8.82
1996	10049.95	11.26
1997	11051.43	12.31
1998	11950.73	39.26
1999	13041.81	30.55
2000	14475.43	28.14
2001	14344.00	27.11
2002	14195.00	27.32
2003	16429.00	29.68
2004	21828.25	34.18

Source: Statistical Bulletin, Department of Tourism and Culture, 2004, Government of India.

Employment Opportunities

Employment opportunity is one of the most remarkable benefits that tourist provides. It is found from the report that 46103 persons were directly employed in tourism sector. More than lakhs of people were employed indirectly in tourism sector every year. Tourism sector is an employment generation sector.

Table 2.8: Employment generated through Tourism in Orissa

Sl. No.	*Sector*	*Number of persons*
1.	Accommodation Sector	7855
2.	Restaurant Sector	14902
3.	Shopping Sector	7325
4.	Transport Sector	11673
5.	Guides Sector	217
6.	Other Sector	4131
	Total Employer Generated	**46103**

Source: Statistical Bulletin, Department of Tourism and Culture, 2004, Government of Orissa.

Increase in tourist inflow is the symbol of economic progress of a state. Tourism creates job opportunities. Employment opportunity is one of the benefits that tourism provides. Table 2.8 gives information on sector-wise employment in tourism in Orissa.

In Orissa there are 860 hotels of different groups for tourists. From among them high spending group of hotels are 69, middle spending group of hotels are 171 and low spending group of hotels are 620. Total led arrangements in there hotels are 33907. In Orissa total 217 recognised guides are working for tourists and also 51 state recognised excursion agencies and 2 Central Government recognised excursion agencies are working (OTDC Report, 2000).

II

VALUATION OF OUTDOOR RECREATION

In order to clearly conceptualise the economic values of natural resources, the concept of Total Economic Value was first defined by Randall and Stoll, (1983). Economic valuation typically focuses on use value in the short run, whether within or outside the market. Ecologists however, are more concerned with ecological values, which provide an underlying long-run nation of value interpreted in a more general sense.

In environmental economics, the Total Economic Value (TEV) of a natural resources such as a wetland ecosystem is considered to comprise of two main sources of value. Use value and non-use value often, option value is added as a third component.

Wet-lands are rich ecosystems, capable of providing a range of goods and services of including recreation to human populations. The value of these goods represent use values. In their discussion of the TEV of tropical wetlands such as mangrove ecosystems, (Aylward and Barbier 1992) distinguishes direct and indirect use values.

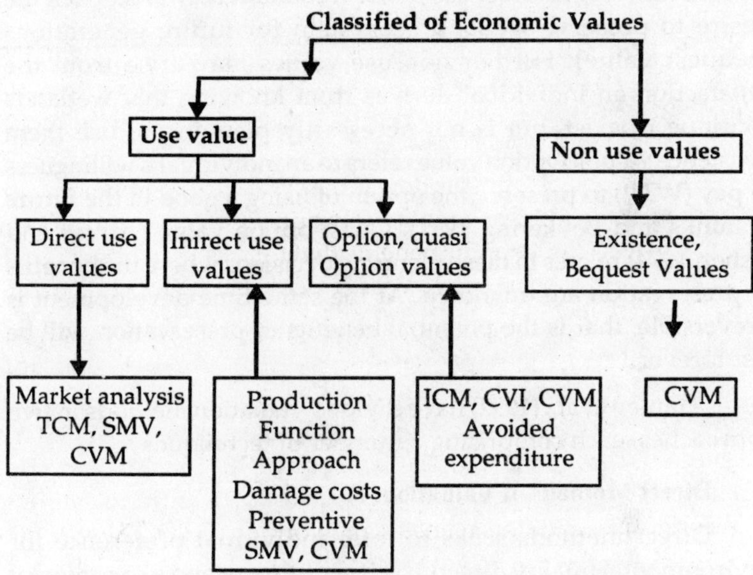

Fig. 2.5: Valuation Methods and Classification

Based on Barbier (1994)

ICM – Individual Choice Models

CVI – Conditional Value of Information Models

CVM – Contingent Valuation Method

TCM – Travel Cost Method

SMV – Surrogate Market Valuation

IOC – Indirect Opportunity Cost Approach

ISA – Indirect Substitute Approach

Direct use values relate to the values derived from direct use or interaction with a wetlands resources and services. Example: Recreation in waterways. Indirect use values stem from the indirect support and protection provided to economic activity and properly by the wetland's natural functions or regulatory environmental services. The classic examples of an indirect use value of wetland ecosystem are flood control, storm protection, habitat/nursery. Non-use-values, on the other hand, are derived neither from current direct or indirect use of the wetland it may arise from the desire to preserve wetland ecosystem for future generations (bequest value). Further non-use values may arise from the satisfaction an individual derives from knowing that wetlands continue to exist, but is not necessarily planning to use them (existence value). Option value refers to an individual's willingness to pay (WTP) to preserve the option of using a good in the future (Paninks and Bevkering 1997). Quasi-option value (Arrow and Fisher, 1974) relates to these planning decisions where the benefits of preservation are unknown. At the same time development is irreversible, that is the potential benefits of preservation will be lost forever.

Hanley et al., (1997) have divided valuation methods in two approaches, each comprising a number of techniques :

1. Direct Methods of Valuation

Direct methods seeks to infer individual preference for environmental quality directly by use of survey and experimental techniques. Contingent valuation (CV) and contingent ranking methods come under this category. This involves asking people for their maximum willingness to pay (WTP) and minimum willingness to accept (WTA) for environmental quality. However, indirect method seeks to elicit preferences from actual observed market based information (Pearce and Moran, 1994). Direct method includes experiments and survey methods.

Experiment Method

In case of experimental method the analyst keeps the respondents in a position in which they can express their hypothetical valuations of real improvements in specific environments (Pearce et al., 1991). In practice it is very difficult to design and implement large-scale experiments, but small-scale experiment can be carried out successfully (Pearce and Moran, 1994).

Survey Method

There are two ways, through which surveys can be undertaken, namely

1. Contingent Ranking Method (CRM)
2. Contingent Valuation Method (CVM)

Contingent Ranking Method

This is similar to CVM. In this method the questioner is content to obtain a ranking of preferences which can latter be anchored by the analyst in relation to a real price of something observed in the market (Pearce and Moran, 1994).

Contingent Valuation Method

The contingent valuation method (CVM) for the valuation of environmental goals was first used by Davis (1963) in a study of hunters in Maine. Later it was developed and become the most widely used and most controversial of all environmental valuation techniques (Hanley et al., 1997). It is used to measure the passive use values of environment resources (Bjornstad and Kahn, 1998, Pearce and Moran, 1994). Under this method people are asked directly in provision of goods or services and their WTA to go for a change or tot rate the change. Here questions are directly asked through a survey to the people to reveal their WTP for some change in a provision or to prevent a change. CVM exercise can be split into five stages:

1. Setting up the contingent or hypothetical market
2. Obtaining bids
3. Estimating mean WTP and /or WTAC
4. Estimating bid curves
5. Aggregating the data

Stage One: The Hypothetical Market

The process of devising a convincing CV scenario consists of a number of elements (Garrod and Willis, 1999). The first step is to set up a hypothetical market for the environmental good that is in the subject of the study. This will include information when the services will be available, how the respondent will be expected to pay for it, how much others will be expected to pay, what institution will be responsible for delivery of the services (Pearce and Moran, 1994). Before main survey, pre-testing should be done using a small focus group. In this case information about all aspects of the hypothetical market is given to the respondents (Hanley et al., 1997).

Stage Two: Obtaining Bids

Once the survey instrument is set up, the survey is administered. This can be done either by face-to-face interviewing, telephone interviewing or mail (Hanley et al., 1997). The face-to-face interview is most appropriate. The NOAA (National Oceanic and Atmospheric Administration) panel came out strongly in favour of this type of interview of CVM. Under this, respondents are presented with a believable vehicle by which the funds will be raised. There are different types of bid vehicles though which WTP bids can be collected such as income tax, value added or sales tax trust fund payment, entry charges, property taxes and utility bills (Garrod, G. and Willis, K.G. 1999, Hanley et al., 1997). Taking WTP as an example, the figure may be derived in several ways :

A. *Bidding game:* higher and higher amounts are suggested to the respondent until their maximum WTP is reached;

B. *Payment card:* A range of values is presented on a card, which helps respondents to calibrate their replies;

C. *Open-ended question:* Here individuals are asked for their maximum WTP with no value being suggested;

D. *Closed-ended referendum:* A single payment is suggested to the respondent, to which he/she either agrees or disagrees (Yes/no reply).

Stage Three: Estimating Average WTP/WTAC

If open-ended, bidding game or payment card approaches have been used, then the calculation of sample mean and/or medium WTP or WTAC is straight forward (Hanley et al., 1997). It is usual in CVM to find that mean WTP exceeds medium WTP, since the former is influenced by a relatively small number of relatively high bids. If a dichotomous choice (DC) method has been used, then the calculation of average WTP/WTAC is more difficult. In this case a suitable econometric model is used for estimation purposes according to the collected data.

Stage Four: Estimating Bid Curves

A bid curve can be estimated for open-ended CVM format, using WTP/WTA as the dependent variable against a range of independent variable. For example – WTP bids might be regressed against income (Y), education (E), age (A) and some variable measuring the quality of environment (Q). The explanatory power of bid curves is considered as a test of the success or failure of a CVM survey (Hanley et al., 1997).

Stage Five: Aggregating Data

Aggregation refers to the process where by the mean bids are converted to a population total value figure. Aggregation of data revolve around 3 issues. First, is the choice of relevant population. Second is moving from the sample mean to the mean for the total population. This can be done by multiplying the sample mean by the number of households, N. Third is the choice of the time period over which benefit should be aggregated.

(H) Indirect Methods of Valuation

Pearce et al. (1991) were of the view that indirect procedure for benefit estimation do not measure direct revealed preference for environmental good. According to Pearce and Moran (1994) the techniques included within the indirect approach are:

- (i) Travel Cost Method
- (ii) Hedonic Price Method
- (iii) Aversive Behaviour
- (iv) Dose-response and Replacement Cost techniques

Travel Cost Method

This technique was first proposed by Hotelling (1931) and popularised by Clawson and Knetsch (1966). Travel cost models are based on an extension of the theory of consumer demand in which special attention is paid to the value of time (Pearce et al., 1991). This approach has relevance for valuing ecotourism (Pearce and Moran, 1994).

The method involves using travel costs as a proxy for the price of visiting outdoor recreational sites. A statistical relationship between observed visits and the cost of visiting is derived and used as a surrogate demand curve from which consumer's surplus per visit-day can be measured.

The basic philosophy of this method is to use the cost of travel as a surrogate for the willingness to pay for using the recreation site. Travel costs include actual transportation cost, fees paid at hotels and at times the opportunity cost of travel time spent on the journey. People staying far from the site will be paying a higher travel cost as compared to those who stay nearby and accordingly, the visitation rate of the former will be smaller than the latter.

The first step in applying the TCM is to collect data through a visitor questionnaire at the site. The list of required data includes transportation expenditure, hotel expenses and park entrance fees, the amount of time spent on travelling, and various socio-economic characteristics of visitors. In the case of economic approaches, the second step is to derive an equation that relates the visitation rate and the independent variables which affect the visitation. Examples of these variables include the total travel cost of visitors and their income, age or other socio-economic characteristics. The third step is to construct a system of demand equations to get an aggregate demand curve for the site. The last step is to measure the area under the aggregate demand curve to get the benefit visitors enjoy from the recreation site.

The travel cost of visiting the resorts may not be easy to capture. The visit to the resort may be combined with other visits and there may be other reasons for travel. For example, the travel cost associated with commuting for those who prefer to live in far away, cleaner areas may be considered expenditure to avoid air pollution. However, the resident may also be paying for a larger house in the suburbs, and not only to avoid pollution. Thus, the TCM valuation procedure also involves statistical complications. Moreover, TCM requires substantial effort to get primary as well as secondary data, and is applicable less to urban degradation than to amenities such as national parks or resorts.

Hedonic Price Method

This was developed by Griliches (1971) and Rosen (1974). It estimates an implicit price for environmental attributes by looking at real markets in which these characteristics are effectively traded (Pearce and Moran, 1994).

Averting Expenditure/Avoided Cost Approach

It assumes complimentarily between market and environment. It takes into account expenditure on private goods, which are used to reduce the environmental degradation.

CONCLUDING REMARKS

Tourism is an exciting activity. It renders immense benefit to the economy by facilitating opportunities for job creation, and contributing to national income. However the natural ecosystem needs to be protected and conserved. Economists have suggested different methods like travel cost method and contingent valuation method to value outdoor recreation.

3

Tourism Potential at Chilika Lake

INTRODUCTION

This chapter speaks about the Lake and her rich biodiversity. A brief discussion is made on the tourism potential at the lake. Attempt is also made to estimate the carrying capacity of the lake. The chapter ends with a brief sample profile and concluding remarks.

THE LAKE AND THE PEOPLE

Chilika lake, situated in the State of Orissa, India, is the biggest brackish waters lake in Asia and a Ramsar site. It is a lake of international importance due to its unique bio-diversity character. This lagoon is situated between latitude 19° 28′ to 19° 54′ north, longitude 85° 51′ to 85° 38′ east covering three districts of the state of Orissa. The lake is pear shaped, with a linear maximum length of 64.3 km and an average mean width of 20.1 km. The mean widths during the summer and monsoon are 14.08 km and 18.0 km., respectively. The water-spread area of the lake varies between 1165 to 906 sq. km. during monsoon and summer. A 35 km. long narrow outer channel connects the main lagoon to the Bay of Bengal near the village Arakhukuda. It is known as Magarmukh channel. This channel has become shallow and thereby obstructs the flow of seawater and free movement of fish to and from the lake. The mouth connecting the channel to the sea is close to the north-eastern end of the lake. High tides near this inlet mouth drive in salt water through the channel during the dry months from December to June, with onset of rains, the rivers falling into the north zone are in spate, causing fresh water currents to gradually push the sea water out. Because of this inlet mouth constantly changes position. Recently the Chilika Development Authority has opened a new mouth near Sipakuda with a hope that the salinity and fish production will increase. Two rivers namely Daya and Vargavi, branches of Mahanadi system along with other small rivulets discharge fresh water directly at the north side of the lake. This creates different varieties of flora and fauna in different parts of the lake, which is responsible for the unique bio-diversity character of the lake Chilika.

Around the lake there are 132 fishing villages with a total population of around 1,04,040. About 26 per cent (27,000) of the village population are active fishermen. Many others depend indirectly on fisheries. They depend on the lake resources for their livelihood. Their socio-economic and cultural patterns have direct bearing on the lake ecology (Mohanty, 1999). The lake is extremely important to local inhabitants and provides fish, fodder and fuel. About 70 per cent of this population depend on fishing as the only means of livelihood. The annual fish catch from the lake is 6000 tons. Fisheries directly support a population of 1,27,000 people, while associated industries and marketing operations supports at least another 50,000.

The complex mix of resources in and around the lake i.e. water, fish, land, forest and fauna, have an inter related effect on community life. People around the lake have the victim of environmental degradation like siltation, weed infestation and water-quality changes. The diminishing resources have affected the socio-economic fabric of the people, their lifestyle, their economic condition, education, health and sanitation, land use pattern and resource use, and resource developmental patterns. All these important issues are to be addressed systematically in an integrated and sustainable manner to protect the lake environment from disaster and to improve the socio-economic status of the people.

Four different communities depend on the lake for their livelihoods in both the direct and indirect way.

(a) the fisherman

(b) the farmers who live around the lake

(c) people depending on forest resources in the lake catchment area for both their livelihood and their fuel requirements.

(d) people depending on tourism.

THE BIODIVERSITY

Chilika lake with unique spatial and temporal salinity gradient is inhabitated by a large biodiversity. It has a unique floral composition. It is home to some rare, vulnerable and endangered species. Total number of fish species vary from 149 to 225 lake's biodiversity, along with a variety of phytoplankton, algae and aquatic plants.

Large number of aquatic birds in the country congregates particularly during the winter. Flocks of migratory waterfowl arrive far from the Caspian sea, Lake Baikal, Aral sea, remote parts of Russia, Central and South East Asia, Ladakh and Himalayas, to feed and breed in its fertile waters. In 1989-90, 2 million birds have visited the lake. As per the report of the Asia waterfowl census, the lagoon supported a million water birds in 1992. The lake hosts over 160 species of birds during the peak migratory season of which at least 97 are inter continental migrants. The lake is also home of irrawaddy dolphins. At present there are at least 50 old irrawady dolphins in the lake.

The shore of lagoon is broadened by green pastures, patches of low-lying vegetation, rocky headlands and promontories. For some distance in the north east of Balugaon, the shore consists of a series of little bags separated by low rocky hills. The southeast end of the lagoon is occupied by two narrow trays separated by rocky hills. Together they from what may be called Rambha basin. This basin is surrounded by hilly areas, the most prominent being the Chantasila hill (Das, 1994).

RECREATIONAL SPOTS

The major attraction of the lagoon for the tourists are its natural beauty with the pleasure of boating, bird watching and cavorting Dolphins while the religious shrine of Kalijai and the mouth-watering delicious dishes are incidental attraction. The important tourist spots of the Chilika are Barkul, Rambha and Satapada.

The followings are some memorable and peaceful excursion spots in Chilika:

Barkul and Rambha

The Orissa State Tourism Development (OSTD) has developed Barkul and Rambha for recreational purposes owing to their scenic beauty. A large swimming pool has been constructed at Barkul. In addition, facilities for water sports like water skiing, scootering, pedal boating, canoeing and rowing are available here. At Rambha, too, the OSTD has developed an area for recreational purposes.

Bird Sanctuary

Nalabana Island has unique features as a habitat for the avifauna. In the island the predominant species among the aquatic plants being *Phragmites Karka*. The Island is a mashy island and covers approximately 32 sq. km. This island is drowned during monsoon with only the reeds and watchtower visible. With the onset of summer the island gradually emerges. At the beginning of the migratory season in October-November, long-legged waters and delving species are predominant. In December-January a large number of ducks, fish eating birds and small waders stop to feed roost. One of the most fascinating sights is the large flocks of flamingos feeding in the shallow waters. During the summer, however only a few residential birds can be seen. Boat is the only communication medium to the islands inside the lagoon. In 1973 the island was designated as the "Chilika bird sanctuary" to provide adequate protection for visiting species.

Birds Island

This is a hillock with huge exposed hanging rocks situated in the southern sector of the lagoon. It is about four kilometres from the Rambha tourist bungalow and one kilometre away on the eastern side of the Ghantasila hill. The island is covered with herb, shrubs, trees and creepers. Numerous resident and migratory avifauna species are also found here. Many of the rocks are painted with bird droppings. Rich algae communities flourish in the waters attract numerous fish species fish. The island is an excellent spot for tourists with specific interest in nature (WISA, 1998).

Badakuda Island (Haneymoon Island)

Situated in Rambha Bay near Sankuda Island, Badakuda Island is 5 km. from the Rambha jetty. It is also known as Honeymoon Island. Water around the island is very clear, dark and blue. The rare and endangered Barakuda burrowing skunk was first recorded by Annandale and Kemp (1915). The island also hosts over hundred species of lowering plants. Some of the bird species are unique to this island.

Breakfast Island

This island is located in the Rambha Bay and between Badakuda and Somolo Island. It is pear shaped also known as Sankuda. A bungalow was constructed on this island by king of Khallikote.

Beacon Island

It is architectural marvel with a conical pillar. It is partially submerged rock on which a small room has been constructed. It is about 3 kilometer from Rambha. As a warning to sailors it was built by Mr. Srodgrass, Collector of Ganjam under the East India Company. The water spread around the Island is very charming.

Bramhapura

It is a large sand bar located between the Bay of Bengal and the lake. The Forest Department has constructed a watchtower and a guesthouse. Tourists visit this area to enjoy the scenic beauty of the sea on one side and birds on the other.

Somolo and Dumkudi

These islands are situated in the central and southern sectors of the lagoon, which are inundated remnants of the Eastern Ghats. Though rocky, they are rich in flora and fauna. Dolphins are often seen in the peripheral water of Somola Island. One can see a beautiful and unique scene of Somolo in the Khallikote village range.

Parikud

A complex of island including Baranikuda, Malatikuda, Badakuda and Sankuda located along the coastal side of the lagoon, was originally known as old Parikuda. These islands are made up of entrenched land and dunes. These are known as the Garh Krishnaprasad Block, ideal spots for nature lovers particularly during the winter season when migratory avifaunas are abundant.

Kalijai Temple

Kalijai temple is situated on an island considered as the abode of the Goddess Kalijai. Domestic tourists are interested in this religious site. Most frequently local boatmen ferry the tourists to visit the temple from Balugaon, INS Chilika and other surrounding areas (WISA, 1998).

Mahayana Temple

Mahayana, temple is dedicated to goddess Mahayana. The site has a natural spring and has appreciable scenic beauty. It is situated 10 km from Rambha (WISA, 1998).

Bhagabati and Dakshya Prajapati Temple

Banpur hosts a shrine of the goddess Bhagabati, Dakshya Prajapati. It is located 13 km from Barkul and 8 km. from Balugaon (WISA, 1998).

Nirmala Jhar

Nirmala Jhar is a perennial stream worshipped as a goddess by the local population. It has significant aesthetic appeal. It is situated 11km. from Rambha and 21km. from Barkul (WISA, 1998).

Palur Canal

The canal has been constructed by the British to improve the movement of waters from the lake to the sea and vice-versa.

Barunkuda

It is a small island situated near Magarmukh with a temple of lord Varuna.

Sand Bar and Mouth of Chilika Lagoon

A beautiful stretch of endless unexplored stretch of empty beach exists across the sandbar, which separates the lagoon from the sea (Chauhan, 1998).

Satapada

One of the main attractions of the lake is the sighting of the Irrawaddy dolphin, once abundant but now an endangered species. The dolphins are often sighted swimming near the mouth of the outlet canal at Satapada and in the central and western parts of the lake.

The Irrawaddy Dolphin (*Oracella brevirostris*)

The Irrawaddy dolphin has a distinguished blunt, beakless head. Females reach a maximum length of about 2 mt to 2.6 mt and a body weight of approximately 190 kg. Males are slightly larger and heavier. The body is streamlined with broad and paddle like flippers. The small dorsal fin is curved with a rounded tip and is placed well behind to body's midpoint. Colour varies form grey to dark blue grey with pallor belly. The mouth has about 17.30 teeth in each raw of the upper jaw and 15.18 in each raw of the lower (WISA, 1998). Named by J.E. Grey from a skull collected by Sir Walter Elliot in the harbour of Vishakhapatnam in the Bay of Bengal, Urcaella is a diminutive of the Latin 'Orce', meaning a type of whale. Brevirostris comes from the Latin brevis on 'short' and restrum meaning 'beak'. Inhabiting the coastal brackish seas and inlets of the tropical and subtrophical Ind-Pacific region, the unusual looking dolphins range from Northern Australia and New Guineu of the Bay of Bengal. They also inhabit some river in the region including the Irrawaddy Mahakum, Mekong, Ganges and Brahmaputra and have been found up to 800 miles upstream (WISA, 1998). Normally, one needs to look for this dolphin in shallow coastal water, particularly turbid estuaries and mangrove swamps.

INFRASTRUCTURE

Infrastructure development is the backbone of the tourism industry. The provision of easy access, clean accommodation, convenient local travel, and opportunities for relaxation and entertainment determine the popularity of a tourist destination. The infrastructure facilities are discussed below.

Mode of Travel

Different ways are there to reach the lake. The most convenient routes are Bhubaneswar to Balugaon, Barkul and Rambha (90 km. 100 km. 130 km, respectively from Bhubaneswar). Satapada near the mouth (50km. from Puri) is yet another convenient gateway to the lake. There are different modes of transport.

- *For tourists travelling by air:* nearest airport is Bhubaneswar. The lake can be approached from different points. The main sites of tourist interest are accessible from Satapada, Barkul and Rambha. Satapada can be approached from Puri, Barkul and Bhubaneswar. One can reach at Rambha from Gopalpur and Berhampur.
- *Second is Rail travel:* One can catch Kolkotta-Chennai route of the South Eastern Railway. It touches the lake at Balugaon, Chilika, Khallikote and Rambha village.
- *Third is the Roadways:* Buses and Taxis to Chilika are available from Cuttack, Bhubaneswar, Puri and Berhampur. Auto-rickshaws are available at Balugaon for Barkul. (WISA, 1998).

To travel inside the lake, both country boats and motor launches are made available both by government and private operators at Barkul, Rambha and Satpada on hire basis. (Table 3.1, 3.2) While in the government service hourly rates differ according to boat capacity, private operators charge according to distance.

Table 3.1: Mechanised motor launch from Barkul, Rambha and Satpada through Government services

Starting point	*No. of Boat*	*Capacity*
Barkul	8	6-34
Rambha	4	8
Satapada	8	15-20

Source: Survey Method.

Table 3.2: Mechanised motor launch from Balugaon, Barkul, Rambha and Satpada through private services

Starting point	*No. of Boat*	*Capacity*
Balugaon	130	20-25
Barkul	70	20-25
Rambha	30	20-25
Satapada	230	20-25

Source: Survey Method.

Accommodation

The Orissa Tourism Development Council (OTDC) has constructed some hotels like Panthanivas at Rambha and Barkul, Yatrinivas at Satapada. There are also same private hotels with A/C facilities. The charges differ from season to season. AC room tariff ranges from Rs. 500 – 800. Non-AC tariff varies from Rs. 150 to 200 (Table 3.3).

Table 3.3: Accommodation status

Place	*No. of Hotels & Panthanivas*	*No. of rooms*		*No. of Beds*	*Room Tariff*	
		AC	*Non-AC*		*AC*	*Non-AC*
Balugaon	3	16	25	79	500-800	200-400
Barkul	2	8	28	72	500-800	300-450
Rambha	1	4	8	30	500-600	200-350
Satapada	1	2	7	18	500-600	150-200

Source: Department of Tourism, Government of Orissa, 2001/Survey.

TOURISM REVENUE

Table 3.4 reveals the information on the tourism earning of OTDC from various sources like restaurant, transport and boating. The maximum earning was from Barkul (17.00 lakhs) in 2003-04. The total earning in the same period is 116.53 lakhs. There are also private restaurants transport and boating facilities. Data is not available relating to their earnings.

Table 3.4: Tourism earnings of OTDC (in lakhs)

Year	*Barkula*	*Rambha*	*Satapada*	*Total*
1998-1999	42.20	7.35	7.97	57.45
1999-2000	25.17	5.17	7.32	37.66
2000-2001	48.79	13.04	15.04	76.87
2001-2002	54.72	16.10	18.41	89.23
2002-2003	62.50	18.49	23.01	104.00
2003-2004	70.53	17.00	29.00	116.53

Source: OTDC (2004), Bhubaneswar.

TOURISTS INFLOW

The maximum tourists arrival to Chilika (Barkul) during 2004 was 2,54,583 where domestic was 2,54,583 and foreign tourists was only 383. During 1994 it was 1,30,86 where domestic tourists was 1,30,625 and foreign tourists was 235 (Table 3.5).

Table 3.5: Tourist arrival (1991 to 2004)

Year		*Ganjam District Rambha*	*Khurda District Barkul*	*Puri District Satapara*
1		*2*	*3*	*4*
1994	D	47304	130625	44000
	F	244	235	103
	T	47548	130860	44503
1995	D	18026	137224	44500
	F	140	150	197
	T	18166	137474	45697
1996	D	20126	155210	47750
	F	130	255	215
	T	20256	155465	47965
1997	D	27322	170830	50620
	F	151	290	224
	T	27473	171120	50844
1998	D	31159	181913	52887
	F	177	319	101
	T	31336	182232	52988

(Contd...)

1		2	3	4
1999	D	29067	154626	41540
	F	75	250	304
	T	29142	154876	41852
2000	D	31723	192940	48492
	F	125	266	183
	T	31848	193166	48675
2001	D	47680	223590	50895
	F	112	285	302
	T	47792	223875	51197
2002	D	63120	250787	54679
	F	268	325	582
	T	63388	251112	55261
2003	D	65800	252060	66262
	F	294	356	587
	T	66094	252416	66849
2004	D	68450	254200	70520
	F	315	383	605
	T	68765	254583	71125

D – Domestic, F – Foreign, T – Total Source: Statical bulletin, Department of Tourism and Culture. Govt. of Orissa (2004).

Tourist arrival to Chilika (Rambha) during 2004 was 68,765 where domestic was 68,480 and foreign was 315.

Tourist arrival to Chilika (Satapada) during 2004 was 71125 where domestic was 70,520 and foreign was 605. Tourist arrived at Satapada gradually increased from 1994. In 1999 the tourists number decreased, again it increased and reached its maximum during 2004.

Table 3.6: Share of tourist arrival to Chilika and Orissa (1994-2004)

Year	*Tourist arrival to Chilika*	*Tourist arrival to Orissa*	*Share of tourist arrival to Chilika (%)*
1994	2,29,725	26,21,016	8.74
1995	2,01,337	27,25,566	7.38
1996	2,23,685	28,07,548	7.96
1997	2,49,437	28,63,213	8.71
1998	2,66,556	28,94,889	9.20
1999	2,25,870	27,17,598	8.31
2000	2,73,689	29,12,115	9.39
2001	3,22,864	31,23,170	10.3
2002	3,69,761	34,36,386	10.7
2003	3,85,359	37,26,270	10.34
2004	3,94,473	41,54,353	9.5

Source: Statistical Bulletin, Department of Tourism and Culture, Government of Orissa (2004).

Total tourist arrived to the lake was maximum during 2004. Share of tourist arrived to the lake was 10.77 during 2002 (Table 3.6 and Fig. 3.1.) *(See Fig. on next page).*

CARRYING CAPACITY OF CHILIKA

The World Tourism Organisation (WTO) reports that "carrying capacity is the level of visitor's use of an area with high levels of satisfaction and minimum impacts on resources" (WTO, 1992). According to Boulding (1985) carrying capacity can be calculated by

$$\text{Carrying capacity} = \frac{\text{Area used by tourist}}{\text{Average individual standards}}$$

Minimum tourist spot area per tourist bed may be used as a standard to measure carrying capacity of beach space.

Three types of carrying capacity are commonly used for analytical purposes (Tantrigama 1998).

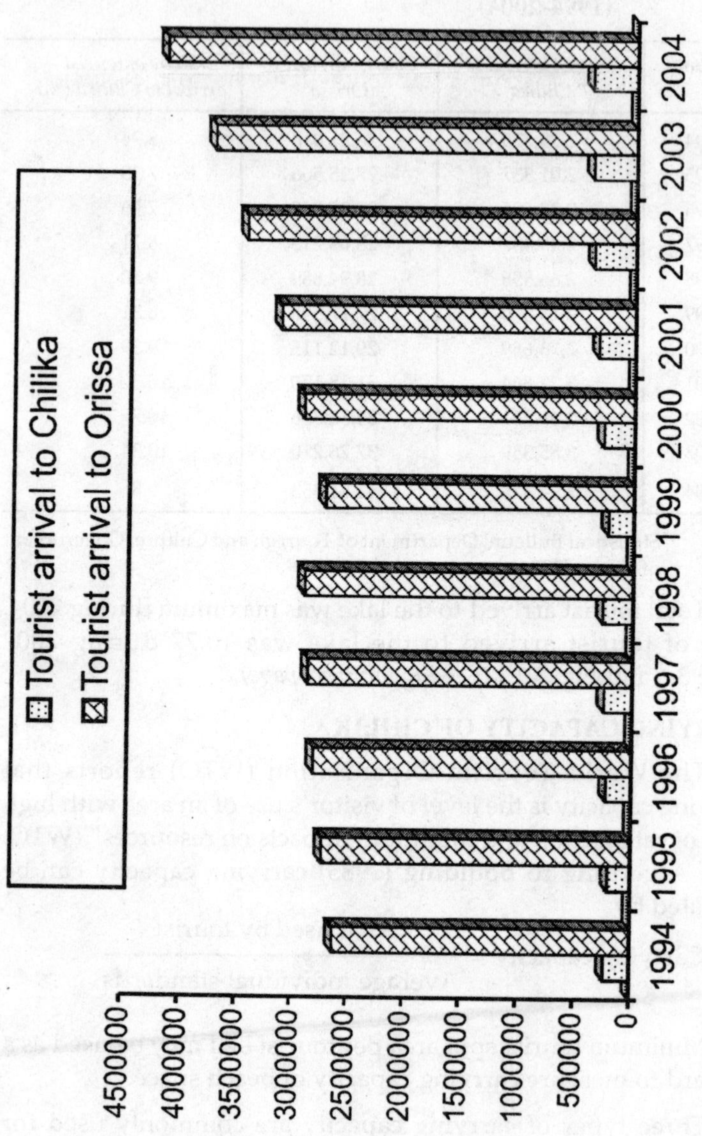

Fig. 3.1: Tourist Arrival to Chilika and Orissa.

(i) Ecological carrying capacity—It refers to the maximum number of tourists that an area can absorb before ecological decline takes place.

(ii) Physical carrying capacity—It refers to the absolute limit on tourist numbers that a resource can cope with. It includes water and electricity facilities, proper waste and sewage disposal, telecommunication and road transport.

(iii) Environmental carrying capacity—It refers to the maximum number of tourists that an area can accommodate without initiating decline in the general perceived attraction of the area. The purpose of estimation of carrying capacity of a particular area is to (i) Minimize the negative impact of tourism (ii) To maintain privacy (iii) To accommodate high budget tourists.

Environmental Carrying Capacity of the Chilika (Barkul)

Environmental carrying capacity of Barkul is estimated as follows.♧

Length of Promenade – 371.00 meter, Width – 192.00 meter

Area of promenade – 371.00 meter x 192.00 meter = 71232 m^2.

Area of tourism office and its surroundings = 4.9 acre

1 acre = 400m^2

Then 4.9 Acre = 4.9 x 4000m^2 = 19600m^2

Length of Panthanivas = 55 meter, width = 39 meter, area of Panthanivas = 55x39 = 2145m^2.

Length of area near Ghata = 22m^2, Width = 10meter, Area = 20 x 10 = 200m^2

So total area used by tourist at Barkul = 71232 + 19600 + 2145 + 200 = 93177 sq.km.

♧ Survey Method.

Table 3.7: Carrying capacity of Barkul during 2001-2004

Sl. No.	*Year*	*Type of carrying capacity*	*Area in square meter*	*No. of tourist arrival*	*Remark*
1.	2001	Environmental carrying capacity	93,177 m^2	2,50,787	Not exceeded
2.	2002	Environmental carrying capacity	93,177 m^2	2,51,122	Not exceeded
3.	2003	Environmental carrying capacity	93,177 m^2	2,52,416	Not exceeded
4.	2004	Environmental carrying capacity	93,177 m^2	2,54,583	Not exceeded

Source: Survey method.

On the basis of W.T.O guideline $50m^2$ space per tourist is required to maintain proper environmental quality. Maximum number of tourist allowable at any point of time is:

$$\frac{93177m^2}{50m^2} = 1864$$

Per year the number of tourist will be 1864 x 365 = 680360. During 2003 and 2004, the total tourist arrival at Barkul were 2.5 lakh and 2.54 lakh. It can be inferred that environmental carrying capacity has not exceeded during 2003 and 2004. Table 3.7 *(See on page 68)* presents the summary output of the carrying capacity of Barkul.

The environmental carrying capacity of Barkul has not reached at a threshold limit. Compared with the existing tourist arrivals at Barkul there is a potential for accommodating more tourists with in carrying capacity limitations. This may be due to the fact that tourist arrivals are not evenly distributed during the peak season or during any day.

ECONOMIC EFFECT OF TOURISM ON LOCAL PEOPLE

Chilika is a tourist spot where above 10000 people depend upon tourism for their living. They directly or indirectly serve tourists. Approximately 500 boat owners depend on tourists because tourists hire-their boat. Betel-shop owners, handicraft shop owners, hoteliers, lodge owners, photographers and other sellers are fully dependent upon tourists. To ascertain the economic impact questions were asked to local shop owners, handicrafts, shop owners, boat owners, betel shopkeeper and hotels. The data collected is analysed below.

Income Generation

It seems from Table 3.8 that September to February and March to August are the peak season for tourists at the lake. The booming tourism promotes hotel business and creates employment opportunities *(See Table 3.8 on next page)*.

Average sale per day of betel shop in September to February is Rs. 450 while daily average sale in March to August is Rs. 150 only. Average sale per day of handicraft shop in September to

February is 550 and in March to August is Rs. 150. Average income per day of boat owners in September to February is Rs. 900. and in March to August it is Rs. 500. These data indicate that sales/income is highest in peak months of September to February and less sale from March to August (Table 3.9).

Table 3.8: Average sale at hotels per day

(in Rs.)

Name of place	*Type*	*No.*	*No. of persons employed*	*Season of business*	
				Sept.-Feb. Sales (in Rs.)	*March-Aug. Sale (in Rs.)*
Barkul	Panthanivas	1	10	800-1000	400-600
	Private hotel	1	5	500-700	400-600
Balugaon	Private hotel	3	20	2000-2200	1500-1700
Rambha	Panthanivas	1	8	700-900	400-600
Satapada	Jatrinivas	1	7	900-1100	400-600

Source: Survey method.

Table 3.9: Average sale from other business

Sl. No	*Name of the establishment*	*No. of Shops/Boats*	*Average sales per day*	
			Sept.-Feb. Sales (in Rs.)	*March-Aug. Sale (in Rs.)*
1.	Betel Shop	30	400-500	100-200
2.	Handicraft	20	500-600	150-200
3.	Boat	460	800-1000	400-600

Source: Survey method.

Employment Generation

It is seen from the Table that 1120 persons have been employed in hotels, betel shops, handicrafts selling and boating. Boating alone employs 1000 persons. Apart from these many persons are employed in transport and other related activities. (Table 3.10).

Table 3.10: Employment generation*

Sl. No	Name of the establishment	No. of persons employed
1.	Hotels	50
2.	Betel Shops	30
3.	Handicrafts	40
4.	Boat	1000
	Total	**1120**

* Excludes employment in transport services and other allied activities related to hotel industry.

Source: Survey method.

SAMPLE PROFILE

Table 3.11 shows income-wise classification of tourists who were interviewed. It is found that the highest 20.91 per cent belonging to monthly income class of Rs. 20,000-25,000, 19.09 per cent belong to the Rs. 15,000-20,000.

Table 3.11: Classification of tourists according to their monthly income

Sl. No.	Income ranges (Rs.)	No. of Tourist	% of total
1.	Below-5000	12	10.91
2.	5001-10000	21	19.09
3.	10001-15000	13	11.82
4.	15001-20000	21	19.09
5.	20001-25000	23	20.91
6.	25001-30000	18	16.36
7.	30001-above	2	1.82
	Total	**110**	**1000**

Source: Sample survey.

Table 3.12 presents tourists according to their age. It is recorded that majority of tourists belong to 36-45 age group. 46 tourists belong to 36-45 age group which constitutes 41.82 per cent of the total. 30.91 per cent belong to 26-35 age group.

Table 3.12: Classification of tourists according to their age group

Sl. No.	Monthly income ranges (Rs.)	Age of tourist in years					
		Upto 25	26-35	36-45	46-55	56-65	Total
1.	Below-5000	3	5	2	1	1	12
2.	5001-10000	1	8	8	2	2	21
3.	10001-15000		4	7	1	1	13
4.	15001-20000		9	10	2		21
5.	20001-25000		4	9	9	1	23
6.	25001-30000		3	9	5	1	18
7.	30001-above		1	1			2
	Total	**4**	**34**	**46**	**20**	**6**	**110**
	% of Total	**3.64**	**30.91**	**41.82**	**18.18**	**5.45**	**100**

Source: Sample Survey.

Table 3.13 highlights the classification of sample tourists on the basis of their educational qualification. 50.91 per cent tourists are graduates and 41.82 per cent are post-graduates and above.

Table 3.13: Classification of tourists according to their level of education

Sl. No.	Monthly income ranges (Rs.)	Level of education of the tourists				Total
		Matriculation or below	Intermediate	Graduate	Post Graduate	
1.	Below-5000	6	1	4	1	12
2.	5001-10000			14	7	21
3.	10001-15000			8	5	13
4.	15001-20000			8	13	21
5.	20001-25000			10	13	23
6.	25001-30000	1		10	7	18
7.	30001-above			2		2
	Total	**7**	**1**	**56**	**46**	**110**
	% of total	**6.36**	**0.91**	**50.91**	**41.82**	**100**

Source: Sample survey.

Table 3.14 gives information about the distance the tourist travelled to reach at the lake. 50.91 per cent of the respondents came from places within 500 kms. from Chilika, less number of tourists came from distance places.

Table 3.14: Classification of tourists according to the range of distance

Sl. No.	Monthly income ranges (Rs.)	*Range of distance (Kilometer)*				Total
		Up to 500	501-1000	1001-1500	1501-2000	
1.	Below – 5000	9	3			12
2.	5001 – 10000	12	5	2	2	21
3.	10001 – 15000	7	4	2		13
4.	15001 – 20000	7	10	2	2	21
5.	20001 – 25000	13	2	6	2	23
6.	25001 – 30000	7	3	5	3	18
7.	30001 – above	1			1	2
	Total	**56**	**27**	**17**	**10**	**110**
	% of total	**50.91**	**24.55**	**15.45**	**9.09**	**100**

Source: Sample survey.

Table 3.15 divided the entire population into family size of upto 4 members and more than 4 members in their families. It is found that the 48.18 per cent of the sample respondents have upto 4 members and 51.82 per cent of the respondents have more than 4 members.

Table 3.15: Classification of tourists according to the family size

Sl. No.	Monthly income ranges (Rs.)	*Level of education of the tourists*		Total
		Up to 4 members	More than 4 members	
1	2	3	4	5
1.	Below – 5000	4	8	12
2.	5001 – 10000	11	10	21
3.	10001 – 15000	8	5	13

(Contd...)

1	2	3	4	5
4.	15001 – 20000	12	9	21
5.	20001 – 25000	9	14	23
6.	25001 – 30000	8	10	18
7.	30001 – above	1	1	2
	Total	**53**	**57**	**110**
	% of total	**48.18**	**51.82**	**100**

Source: Sample survey.

91.82 per cent of the tourists are male and only 8.18 per cent are female. In most cases, the families were accompanied by male members. They often responded to the survey questionnaire (Table 3.16).

Table 3.16: Classification of tourists according to the sex

Sl. No.	*Monthly income ranges (Rs.)*	*Male*	*Female*	*Total*
1.	Below-5000	5	7	12
2.	5001-10000	20	1	21
3.	10001-15000	13		13
4.	15001-20000	21		21
5.	20001-25000	22	1	23
6.	25001-30000	18		18
7.	30001-above	2		2
	Total	**101**	**9**	**110**
	% of total	**9.82**	**8.18**	**100**

Source: Sample survey.

Table 3.17 shows that most of the tourists stay for one or 2 days at Chilika. Even tourists coming from outside Orissa use to stay for 10 days either at Bhubaneswar [illegible] at Puri. They stay at Chilika for 1 to 2 days only.

Table 3.17: Days spent by Tourists stay at Chilika

Sl. No.	*Monthly income ranges (Rs.)*	*No of days staying inside Chilika*		
		Up to 2 days	*More than 2 days*	*Total*
1.	Below – 5000	12		9
2.	5001 – 10000	18	3	21
3.	10001 – 15000	12	1	13
4.	15001 – 20000	19	2	21
5.	20001 – 25000	23		23
6.	25001 – 30000	17	1	18
7.	30001 – above	2		2
	Total	**103**	**7**	**110**
	% of total	**93.64**	**6.36**	**100**

Source: Survey method.

Table 3.18 highlights the number of tourists came from different parts of the India. Out of 110 samples 40 tourists are from Orissa followed by West Bengal, Andhra Pradesh, Delhi, U.P. and Jharkhand, Bihar, Kerala & Karnataka, Maharastra, Meghalaya, Tamil Nadu, Pandichery, Punjab, Assam and Goa.

Table 3.18: No. of tourists visit from different states

Sl. No.	*State*	*No. of tourist*
1	2	3
1.	Orissa	40
2.	Delhi	6
3.	Bihar	3
4.	Andhra Pradesh	16
5.	Meghalaya	1
6.	Uttar Pradesh	5
7.	Pandicherry	1
8.	Kerala	2
9.	West Bengal	23

(Contd...)

1	2	3
10.	Jharkhand	5
11.	Karnataka	2
12.	Maharashtra	2
13.	Tamil Nadu	1
14.	Punjab	1
15.	Assam	1
16.	Goa	1
	Total	**110**

Source: Survey method.

Opinion and Suggestions of the Tourists

All the respondents feel that the infrastructure facilities such as rest shed, lavatory, good hotels, restaurant, transportation and communication, boating, parking place and transportation available at Chilika lake are inadequate which need to be developed. 32 per cent of tourist told that there was high boating cost, where as 22.73 per cent opine in favour of providing lifejackets in boat. 58.19 per cent of tourists are not satisfied with the bird watching they suggested to make some bird watching towers. 58.19 per cent respondents feel the need for expert guide. 22.73 per cent of tourists feel the need for swimming. 40 per cent of respondents argue to develop off-season tourism. 57.27 per cent of tourist focussed on proper information and advertisement about the lake. 67.27 per cent tourists are not satisfied with the method of waste disposal. 61 per cent of respondents feel the need for eco-tourism and awareness of both the local people and tourists (Table 3.19).

Table 3.19: Opinion and suggestion of Tourist

Sl. No.	*Types of opinion and suggestions*	*Response No.*	*%*
1	2	3	4
1.	Infrastructure facilities are quite inadequate, it needs to be developed, including boating	110	100
2.	High boating cost	36	32.73

(Contd...)

1	2	3	4
3.	Provide lifejackets in boat	25	22.73
4.	No proper bird watching and need watching	64	58.19
5.	Expert tourist guide is necessary for identifying the birds and species	64	58.19
6.	Develop underwater swimming	25	22.73
7.	Develop off season tourism and make organised and professional	44	40.00
8.	Provide proper information and advertisement	63	57.27
9.	Clean the environment and disposal of the waste	74	67.27
10.	Promote eco-tourism and awareness of the people	68	61.82

Source: Opinion survey.

CONCLUDING REMARKS

Chilika lake is the largest lagoon in Asia, one among six Indian wetlands declared under the Ramsar convention in 1982. It is a rich preserve of ecological diversity with over four hundred vertebrates of both the brackish and fresh water species. More than a million migratory waterbowls and shore birds gather here during winter. Several among them are recorded as endangered and threatened or vulnerable species. It is a very large lake with a drainage basin of over 4,300 sq. km. It is, therefore, a fertile fishing ground and a lifeline for over 1 lakh fishermen. Though attracting tourism and export of prawn crab and fish, it contributes to India's foreign exchange balance.

4

Data Analysis

INTRODUCTION

This chapter is a modest attempt to estimate the total recreation value (TRV) of Chilika lake. The TRV has been estimated by using both CVM and TCM. For the purpose of estimation the data structure required some modification and reconstruction. Out of 110 visitors, we have taken 100 observation, we have excluded 10 observations, as these are the extreme cases. The name, definition and descriptive statistics of variable are given in Table 4.1.

Table 4.1: Definition of the variables

Variables	–	*Description*
WTP	–	Willingness to Pay (in Rs.)
RDISTN	–	Respondents Distance to Chilika in (km.)
RINBID	–	Respondents initial bid amount (in Rs.)
BOATRE	–	Boating expenditure of the respondent (Rs.)
HINCOM	–	Household monthly income (Rs.)
FASIZE	–	Family size of the respondent (No.)
VISAGE	–	Visitor's age (years)
VISEDU	–	Visitor's educational qualification (No. of years)
NOSITE	–	No. of sites visited by the respondent (No.)
NUMVIT	–	No. of visit to the lake (No.)
SDUMMY	–	Sex dummy (male = 1, female = 0)
LDUMMY	–	Lake quality dummy (good = 1, otherwise = 0)
VPRICE	–	RTTEXP + RTVALU= (in Rs.)
RTTEXP	–	Respondents total travel expenditure (in Rs.)
RTVALUE	–	Wage of no. of days visited, (per day minimum wage Rs. 60)

ESTIMATION OF RECREATION VALUE THROUGH CV METHOD

The Models

Two different types of regression models are estimated for the WTP elicited from the sample respondent i.e. linear regression and non-linear regression. The first set of models use OLS for determining linear regression relationship. The second type of models use non-linear probability model due to the dichotomous structure of the dependent variable. Both the forms of WTP are considered as dependent variables with binary choice (1.0). After making trials with all possible combinations of explanatory variables, the following seven variables are chosen, RDISTN, RINBID, BOATRE, HINCOM, FASIZE, VISEDU, NOSITE, NUMUIT.

Linear Regression Model

The linear regression model takes the following form:

$$y_i = \alpha + \beta_j X_{ij} + e_i \tag{1}$$

Where, y_i is dependent variable for household I (I = 1, 2, ...100), X_{ij} is the value of the jth explanatory variable of household i. α is the intercept. There are several explanatory variables and β_j is the slope coefficient with respect to the jth explanatory variable and e_i refers to the error form.

Binary Choice Models

In binary choice model y is a dichotomous choice variable,

Y = 1 (yes WTP)

Y = 0 (No WTP)

For this dichotomous choice as the dependent variable both logit and probit models have been tried. The models are estimated with maximum likelihood estimators (MLE).

Probit Model

The probit probability model is associated with the cumulative normal probability function. To understand this model, assume that there exists a theoretical continuous index Z_i which is determined by an explanatory variable X. The functional form of the probit model is (Pindyck and Rubinfeld, 2003 and 2004).

$$Z_i = \alpha + \beta X_i \qquad (2)$$

Observations on Z_i are not available. Instead, we have data that distinguish only whether individual observations are in one category (high values of Z_i) or a second category (low values of Z_i). Probit analysis solves the problem of low to obtain estimates for the parameters α and β while at the same time obtaining information about the underlying index Z.

The standardize cumulative normal function is written

$$P_i = F(Z_i) = \frac{1}{\sqrt{2\Pi}} \int_{-\infty}^{Zi} e^{-s^2/2} \, ds \qquad (3)$$

where s is a random variable which is normally distributed with mean zero and unit variance. By construction, the variable P_i will lie in the (0, 1) interval. P_i represents the probability that an event occurs. Since this probability is measured by the area under the standard normal curve from -∞ to Z_i, the event will be more likely to occur the larger the value of the index Z_i.

Logit Model

The Linear Probability Model (LPM) is in the form of

$$Pi = E(Y = 1/X_i) = \beta_1 + \beta_2 X_2 \qquad (4)$$

Where X_i is represents the explanatory variables of visitor and Y = 1 means the yes WTP and Y = 0 means the no WTP. The logistic model is based on the cumulative logistic probability function and is specified as (Gujarati, 2004, p 595).

$$P_i = E(Y = 1/X_i) = \frac{1}{1 + e - (\beta_1 + \beta_2 Xi)} \qquad (5)$$

where e is the familiar base of the natural logarithm. For ease of exposition we write

$$P_i = \frac{1}{1 + e^{-Z_i}} \tag{6}$$

where $Z_i = \beta_1 + \beta_2 X_i$

Equation (6) represents the cumulative logistic distribution function.

As we know P_i ranges from 0 to 1 so it is easy to verify that Z_i ranges from -∞ to ∞ and that Pc is non-linearly related to Z_i (i. E,- X_i). There is estimation problem because P_i is not only non-linear in X but in β's as well as seen equation (5). This means that we can not use the familiar OLS procedure to estimate this parameters. But this problem is more apparent that real because equation (5) is intrinsically linear which can be shown as follow:

If P_i, the probability of yes for WTP, is as given by equation (6), then $(1-P_i)$, the probability of no for WTP is:

$$1 - P_i = \frac{1}{1+e^{Z_i}} \tag{7}$$

Therefore, we can write

$$\frac{P_i}{1-P_i} = \frac{1 + e^{Z_i}}{1 + e^{-Z_i}} = e^{Z_i} \tag{8}$$

Now $P_i/1-P_i$ is simply the odds ratio in favour of willingness to pay – the ratio of probability that a respondent say yes for WTP to the probability that is no for WTP.

Now if we take the natural log of equation (8), we obtain

$$L_i = \ln = \left(\frac{P_i}{1-P_i}\right) = \ln(e^{Z_i}) = Z_i = \beta_1 + \beta_2 X_i \tag{9}$$

that is, L, the log of the odds ratio, is not only linear in X but linear in the parameter also. L is called the Logit and hence equation (9) is the Logit model.

The Results

In Table 4.2 the co-efficient of the determinants of the WTP of the respondents are given in linear regression models. The estimated model shows significant result. The t-ratio is significant for all the variables excluding visitor's education. Also in these models the adjusted coefficient of determination ($\bar{R}^2$) is good. All the variables except VISEDU positive influence on the WTP. But in case of qualitative dependent variables the non-linear regression models are more preferable. So for WTP estimation, discrete choice models like logit and probit have been used. The outcome of these models are analysed in Tables (4.2 and 4.3).

Table 4.2: Influence of the explanatory variables on WTP in linear regression models

Sl. No.	*Intercept & explanatory variables*	*Coefficient (t-ratio)*
1.	Intercept	0.5364 (1.858)*
2.	RDISTN	0.0001950 (2.841)***
3.	RINBID	0.0002739 (3.223)***
4.	BOATRE	0.0006566 (2.006)**
5.	HINCOM	7.9816E-006 (2.657)***
6.	FASIZE	0.0601 (1.993)**
7.	VISEDU	- 0.0222 (- 1.479)
8.	NOSITE	0.1475 (4.021)***
9.	NUMVIT	0.2101 (6.859)***
10.	$\bar{R}$	0.542

Table 4.3 presents estimated logit and probit models for WTP. t-ratio are significant in case of logit model (logit and probit). There is no consistency between the results of both the models. However, a comparison of the estimated models shows that the WTP logit model is best in terms of the t-ratio. The estimated equation, which is an index function for probability, is

$\hat{P}$ (WTP = 1) = –6.9647 + 0.0012 RDISTN + 0.0097 RINBID + 0.0055 BOATRE

(- 2.073) (1.905) (2.005) (1.997)

+ 0.0000751 HINCOM + 0.6379 FASIZE – 0.4122 VISEDU

(2.022) (1.806) (- 1.981)

+ 1.1626 NOSITE + 2.7726 NUMVIT (10)

(2.094) (2.348)

Moreover, the model correctly predicted 96 per cent of probabilities. The coefficients in the above equations and the marginal effects estimated through the Gretl Software. All these suggest that the probability of saying "yes" to the WTP question is positively influenced by all the variable except VISEDU. The negative influence of VISEDU might be indicative of the indifferent attitude of the highly educated tourist towards the protection of Chilika.

Table 4.3: Determinants of WTP in logit and probit models

Sl. No.	*Intercept & variables*	*Coefficient (t-ratio)*	
		Logit	*Probit*
1.	Intercept	6.9647 (2.073)	4.3777 (1.954)
2.	RDISTN	0.0012 (1.905)	0.0006770 (1.867)
3.	RINBID	0.0097 (2.005)	0.0116 (2.596)
4.	BOATRE	0.0055 (1.997)	0.0032 (1.813)
5.	HINCOM	0.0000751 (2.022)	0.0000446 (1.935)
6.	FASIZE	0.6379 (1.906)	0.4093 (1.880)
7.	VISEDU	- 0.4122 (- 1.981)	- 0.2599 (- 1.795)
8.	NOSITE	1.1626 (2.094)	0.6333 (1.720)
9.	NUMVIT	2.7726 (2.348)	1.8556 (1.686)

Total Recreational Value of Chilika Lake Using CV Method

Estimation of the consumer surplus is necessary to calculate the recreational value of Chilika lake. The first step is to estimate a logit function which models probability of a positive response to willingness-to-pay. The function was specified generally as:

$$Y = \frac{1}{1+e^{-(\beta_0+\beta_1 X_i)}} \quad (11)$$

where Y = 1 if a respondent answered "no", X_i is the vector of quality and preference variables, and β_0 and β_1 are parameters to be estimated. Equation (11) can be used to derive a bid function for wetland-based recreation. This bid function can then be used to derive a demand function for wetland based recreation (Sellar, Chavas and Stall, 1986). Mean willingness-to-pay for wetland based recreation can be estimated for this demand function.

The mean willingness-to-pay for recreation is Rs. 198 per visitor. It is very low. In valuation literature an alternative way is suggested to estimate the WTP. The WTP can be derived dividing the summation of the constant term and all the slope coefficients in the model (other than the bid amount) by the coefficient of the bid amount variable (Loomis and caban, 1998, pp. 315-322, Pate and Loomis, 1997, pp. 199-207; Langford and Batemen, 1993, pp. 1-29).

Thus adopting this method the expected value of WTP is

$$E(WTP) = \frac{\sum \beta_{i-1}}{\beta_2} \tag{12}$$

where β_{i-1} refer to the sum of the intercept and all the slope coefficient except bid amount, and β_1 refers to the coefficient of bid amount variable. This expected WTP is estimated at Rs. 1148, which is higher than the mean willingness-to-pay stated earlier.

In order to estimate the total recreational value of Chilika we can multiply the average number of visitors for the period of 1994 to 2004. Since the sample in this study covers only domestic visitors we calculate the recreational value of the domestic tourist only. The average number of total tourist is calculated as 1,07,173. This number is to be multiplied by the individual consumer surplus (CS) Rs. 1148 Thus the TRV can be estimated as Rs.123 million per annum.

ESTIMATION OF RECREATION VALUE THROUGH TRAVEL COST (TC) METHOD

In order to model the travel cost function, we follow Freeman (1993) and assume that the individual's quality depends on the total time spent at the site, and the quality of the lake. With the duration of the visit, for simplicity, the time on site can be represented by the number of visit.

In the travel cost models, the consumer is assumed to choose his/her visit to a lake (v), just as choose a basket of other goods (x). Let P_o be the cash – costs associated with a visit to the lake and assume x to be the enumerative. The lake has a level of quality (or anything else that affects the quality of a visit to the lake). Higher q_s better we are interested in visit over the course of a fixed period of time, say a year. Assume that units in which x is measured are such that the price of x is unity. Let P_o be the cash-cost expenses associated with a single trip to the park – automobile, train or plane expenses, accommodation and admission charges. Also assume that the individual works L hours at a wage w. The utility maximization problem is solved

$$\max U(x,v.q) \quad (1)$$

x.v

Subject to the budget constraint

$$wL = x+PoV \quad (2)$$

Consider now the additional costs to be borne by the individual in visiting the lake. Time must be spent to travel and to visit the lake. Consider the individual has T hours of time to spent on working or visiting lake. If t_t is the travel time to and from the lake and t_v is the time spent at the lake, the individual faces the time constraint:

$$T = L+(t_t+t_v)V. \quad (3)$$

Equation (3) can be substituted for Equation (2) and the maximization problem can be re-written as:

$$\max U(x.v.q) \quad (4)$$

x.v

Subject to

$$wT = x+[P_o+W(t_t+t_v)]v$$

$$= x+P_v v \quad (5)$$

Here Travel Cost (TC) is given by the price of the visit:

$$TC = P_o + w(t_t + t_v) = P_v \quad (6)$$

This utility maximization problem is conventional except that the "price" of a visit considers not only money spent but also value of time devoted to the visit, with time valued at the wage rate. Solving the above maximization problem will yield a demand function for visit to the lake for each individual i:

$$Vi = f(tV, XS, \ldots\ldots\ldots\ldots, XE) \quad (7)$$

This would form the basic of an individual travel cost model. Here XS,.........XE are other socio-economic variables such as income, education, age, preferences and close substitutes that may be included in the model (Hanley and Spash 1993). The individual travel cost method uses individual data separately but requires that visitors should visit the lake frequently in any one year.

The Models

The following functional forms have been statistically tested

(1) Linear $r = \alpha + \beta P$

(2) Log – Linear $\log r = \alpha + \beta P$

(3) Double – log $\log r = \alpha + \log \beta P$

(4) Negative exponential $r = \alpha + \log \beta P$

The estimated consumer surplus for an individual making r visit to the site in case of a linear form is given by $CS = r^2/2\beta$. The linear functional form implies finite visits at zero cost and has a critical cost above which the model predicts negative visits. The consumers surplus in case of the log-linear functional form is given by $CS = -r/\beta$. The log-linear functional form has been widely used in travel cost models. It implies a finite number of visits at a zero cost and never predicts negative visit, even at a very high cost (Garrod and Willis, 1999).

The basic model used in this study depicts the numbers of visits to Chilika lake. It is function of factors such as the travel cost, time spent in travelling, income, age, education, family size, site quality, number of sites in the lake etc. Thus the model may be specified as follows:

r_i = Bo+B_1 travel cost + B_2 household income + B_3 age of visitor +B_4 visitors highest level of education + B_5 household size + B_6 D1 (lake quality dummy)+ B_7 number of sites in the lake + e_i (8)

Where r_i, the dependent variable, stands for the number of visits by the i ith individual to Chilika lake per period of time, travel cost means round trip total cost from an individual's residence to and from the site and includes the opportunity cost of travel time and stay at lake. D_1 = 1 if the visitor's perception about the site's recreational facilities is good and O otherwise.

Results

We have tried all possible combinations of explanatory variables, on the above four functional forms. The following six variables are chosen for our analysis.

NUMVIT, VISAGE, HINCOM, VISEDU, LDUMMY, NOSITE

Out of the four models the log-linear model shows significant result. The t-ratio is significant for all the variables. The adjusted coefficient of determination (R^2) is good. In case of linear model except income all variables are significant. Similarly both the double log and negative exponential model the intercepts are not significant. The log-linear models is the best-fit model (Table 4.4). The estimated equation is:

NUMVIT (V) = 1.1808 - 0.0001039 VPRICE + 0.0091 VISAGE

(3.051) (-4.562) (1.906)

+ 5.13e $^{-6}$ HINCOM - 0.0428VISEDU + 0.2764 LDUMMY

(1.993) (-1.980) (2.822)

- 0.1369 NOSITE (13)

(-2.928)

So far as the qualitative significance of the coefficients of the explanatory variables are concerned the results are found as per normal expectation. The travel cost exhibits a negative sign in all the models. This implies that as in the standard demand theory, quantity demanded has an inverse relationship with its price. All the models further reveal that the marginal effect of income on travel cost demand for Chilika is positive. The other variables like age and lake quality have positive influence on the recreation demand. Educations (VISEDU) have negative influence on the recreation demand. The NOSITE also negatively influence on the number of visit. It seems that the number of sites have negative influence on the number of trips. It may be due to the fact that visit to increased number of sites tend to increase the case of trip.

Table 4.4: Determinants of number of visits

Sl. No.	Intercept & variables	Co-efficient (t-ratio)			
		Linear model	*Long-linear model*	*Double log model*	*Negative exponential model*
1.	Intercept	3.5618	1.1808	1.5783	4.2123
		(3.943)	(3.051)	(1.290)	(1.447)
2.	VPRICE	-0.0002015	-0.0001039	-0.3627	-0.7236
		(-3.791)	(-4.562)	(-6.609)	(-5.540)
3.	VISAGE	0.0295	0.0091	0.2161	0.7952
		(2.095)	(1.906)	(0.974)	(1.506)
4.	HINCOM	9.7069e^{-6}	5.13e-6	0.2601	0.5156
		(1.586)	(1.993)	(2.702)	(2.251)
5.	VISEDU	-0.1137	-0.0428	-0.5705	-1.5332
		(-2.251)	(-1.980)	(-1.931)	(-2.181)
6.	LDUMMY	0.557	0.2764	0.2356	0.4723
		(2.431)	(2.822)	(2.608)	(2.197)
7.	NOSITE	-0.3495	-0.1369	-0.3534	-0.9866
		(-0.203)	(-2.928)	(-2.750)	(-3.226)

In the above functional form the price or travel cost (VPRICE) elasticity at the point of the mean is estimated at – 0.1242. This estimated elasticity has usual interpretation that 1 per cent change in total travel cost will lead, on an average, to approximately a 0.12 per cent decrease in the number of visit to Chilika for recreation. Similarly, the estimated income elasticity of demand at the point of mean is 0.06712. When the scaling of income in thousands of Rupees is adjusted the elasticity comes to 67.12 which is higher than 1. Thus, we could classify the recreation in Chilika as, a luxury rather than a necessity, which is consistent with what one would expect for an average household in India in general and in Orissa in particular.

Total Recreational Value of Chilika-TC Method

The prime job is to estimate the consumer surplus (CS). The consumer surplus estimates vary considerably depending on the type of TCM used and the specification of functional form specified. The appropriate measure of consumer surplus is the area under the demand curve i.e. between 0 and r (r=number of visit). Once CS is estimated per individual the aggregate CS can be calculated by multiplying individual CS with the estimated average number of individuals visiting Chilika.

Table 4.5: Consumer Surplus per visitor

	Linear model $Cs=-r^2/2\beta$	*Log linear Model* $CS = -r/\beta$
Using total travel cost variable. Set A	13552.57	9566.52
Using local cost variable. Set B	1700	1200

In the above table set A is taken in account the number of visit and β is taken as total travel cost variable. Similarly in set B r represent the number of visit and β is only local cost variable. In Table 5.4 where r refers to the number of visit and β refers to the travel cost coefficient in the total travel cost variable and local cost variable. Since the log linear functional form has been widely used in travel cost models, we use the local cost variable in log linear model for calculating the CS. Applying the formula to stated model the consumer surplus for individual visitor is estimated at Rs. 1200.00.

In order to estimate the total recreational value of Chilika we have multiplied the average number of visitors (for the period of 1994 to 2004) by individual CS. For estimation of individual CS in our log linear model, we have only taken into account the local cost variable. Thus the CS for individual visitor is estimated at

Rs. 1200. The average number of tourists is as 107173. This number is multiplied by the individual consumer surplus (CS) of Rs. 1200.00. Thus the TRV of Chilika lake is estimated at 128 million per annum.

CONCLUSION

In this chapter the collected data has been processed to measure consumer surplus associated with on-site recreation uses of Chilika Lake. Aggregate consumer's surplus was estimated both through contingent valuation (CV) method and Travel Cost (TC) method. In both the methods the estimated consumer's surplus was Rs. 1148 and Rs.1200 respectively per annum. Thus yields an aggregate gross economic value estimate of about Rs. 123 million and Rs. 128 million annually respectively.

5

Conclusion

It is the time to tie up the work. Conclusions are drawn on basis of the research findings interred in the previous chapters. Recommendations are made accordingly.

SUMMARY OF THE FINDINGS

Tourism involves a temporary break with normal routine. It is a pleasure activity. It relates closely with the location of the tourist spot and mobility of the people from place to place. It has been recognised as one of significant component of the tertiary sector, contributes in a major way to the national income. Probably it is due to these facts, countries both developed and developing, started emphasising on tourism development (Chapter 2). Tourism provides an ample job opportunities to people. It enhances scope for self-employment also. It generates job opportunities for millions in the formal as well as in the informal sector. The state of Orissa is no exception to it. Natural wetlands provide many important functions for humankind besides tourism. Economic valuation methods offer a more comprehensive assessment of the goods and services provided by wetland ecosystems, and hence may contribute to more informed decision-making (Chapter 2).

The present study focuses mainly on economics of tourism and recreation value of Chilika lake. Literatures are browsed and research is developed around this problem. The study attempts to estimate the recreation value of the lake using both the TCM and CVM method. The objectives, methodologies and hypotheses are clearly spelt out (Chapter 1). The primary data has been collected from hundred households through a sample survey. A specific designed questionnaire has been prepared as an instrument of data collection. The data have been processed, tabulated and analysed to present results. Econometric tools like regression analysis and logit, and probit models have been used (Chapter 1).

Chilika lake situated in the state of Orissa, is the largest brackish water lake in Asia. It is a lake of international importance because of her unique beauty and precious biodiversity. The lagoon is a home for hundreds of species of fish, prawn and crab. It provides livelihood to thousands of fishermen. In the lake there is the religious shrine of Kalijai on a tiny island. The lake is thus an attraction for the tourists, domestic as well as foreign, and nature lovers (Chapter 3).

To estimate the recreational value of Chilika lake we have used both the Travel Cost Method and Contingent Valuation Method. In case of CV method the expected WTP for recreation is estimated at Rs.1148 per visitor. The total recreational value of Chilika is estimated at Rs. 123 million per annum. In TCM to model the travel cost function Freeman (1993) has been followed. The basic model used in this work depicts the number of visits to the lake. Four functional forms are tried with possible combinations of explanatory variables. Out of the four models the log-linear model shows significant result. It is found in this best fit model that travel cost exhibits a negative sign. It means as the cost of travel increases the demand for visit or number of visit decreases. The total recreational value of the lake is estimated of Rs. 128 million per annum.

VERIFICATION OF THE HYPOTHESES

It is now the time to confirm or falsify the hypotheses.

Chilika is the largest brackish water lagoon, one among six Indian wetlands declared under the Ramsar convention in 1982. It is a rich preserve of ecological diversity with over four hundred vertebrate is of both the brackish and fresh water species. More than a million migratory waterfowls and shore birds gather her during winter, several among them recorded as endangered and threatened or vulnerable species.

Several small islands like Nalabana, designated as a bird sanctuary since 1973, Kalijai, Somala, Dukudi, Honeyman, Breakfast and Bird island can be seen in the lake. The avifauna of the lake consists of 150 species of birds. Dowitcher, one of the least known Asian shorebirds and the spoonbill sandpiper, among the nearest are some of the interesting bird fauna of the lake. All these make the lake an attractive spot for tourists. These finding confirm the *HYPOTHESIS I*.

In chapter 4 we have tried all possible combinations of explanatory variables using the four functional forms. Out of the four models the log-linear model seems to be the best – fit. The results in this model are as per our expectation. The travel cost exhibits a negative sign. It implies that as in the standard demand theory, quantity demanded has an inverse relationship with its price. These finding confirm *HYPOTHESIS II*.

The attitude towards protection and conservation of the lake is assessed using the contingent valuation method of survey of tourists. This can be stated as WTP for conserving the lake. As per our expectation the findings of regression analysis (Table 4.3, Chapter 4) reveal that a rise in income of tourist tend to influence the WTP of tourists. *HYPOTHESIS III* is thus confirmed.

The recreation value of the lake has been estimated at Rs. 123 million (CVM) and Rs.128 million (TCM). The recreation value of the lake is high enough to justify protection and conservation of the lake. This confirms the *HYPOTHESIS IV*.

THE FUTURE ACTION

The above conclusions pave the path for suggesting future course of actions to protect the lagoon.

The Chilika Development Authority (CDA) can play an important role in promoting eco-tourism in the lake.

There should be strong commitment on the part of the stakeholders: policy makers, government, tourists and people to develop and integrate ecotourism with aqua-cultural promotion. Such an integrated approach is possible only if the authority such as CDA arms itself with all these tasks together.

Preservation of the lake should not be perceived as a policing exercise. Rather the village communities should be involved in all major tasks such as promoting eco-tourism, monitoring, policing, de-weeding, technological choices and secondary employment generation.

Social inefficiency exists in the sustainable management of the lake fishery resources. Several failures like Natural lake use conflicts, fishermen and non-fishermen conflicts, market failures and intervention failures should be carefully corrected.

The total recreation value of Chilika is estimated at about more then Rs. 1 crore per annum is quite significant. This should justify very high investment for conservation and all round development of the lake.

Bibliography

Alberine, A., Boyle, K. and M. Welsh (2003), "Analysis of Contingent Valuation Data with Multiple Bids and Response Options Allowing Respondents to Express Uncertainty", *Journal of Environmental Economics and Management*, 45(1), pp. 40-62.

Andronikou, A. (1987), "Cyprus: Management of the Tourist Sector", *Tourism Management*, Vol. 7(2) pp. 127-129.

Anita, P. (1997), *Tourism Impact*, Foundation Course in Tourism, TS, 1-9, IGNOU, New Delhi.

Annandole, W. and S. Kemp (1915), *Introduction, Fauna of the Chilika Lake*, Mem Ind. Mis(1), pp. 1-20.

Archer, B.H. (1976), "Demand Forecasting in Tourism", *Occasional Papers in Economics*, No. 9, Bangor: University of Wales Press.

Archer, B.H. (1977), *Tourism Multiplies: The State of the Art*, Occasional Paper in Economics, No. 11, Bangor: University of Wales Press.

Archer, B.H. (1995), "Importance of Tourism for the Economy of Bermuda", *Annals of Tourism Research*, Vol. 22(4), pp. 918-930.

Archer, B.H. and C. Owen (1971), "Towards a Tourist Regional Multiplier", *Regional Studies*, Vol. 5(4), pp. 918-930.

Archer, B.H. and J. Fletcher (1996), "The Economic Impact of Tourism in the Seychelles", *Annals of Tourism Research*, Vol. 23(1), pp. 32-47.

Arrow, K.J. and Fistor, A.C. (1974), "Environmental Preservation, Uncertainty and Irreversibility", *Quarterly Journal of Economics*, 88, pp. 312-319.

Attanayake, A., Samaranayak, H.M.S. and N. Ratnapala (1983) "Sri Lanka, in E.A. Pye and T.B. Lin (eds.), *Tourism in Asia: The Economic Impact*, Singapore; Singapore University Press.

Aylwand, B.A. and E.B. Barbier (1992), "Valuing Environmental Functions in Developing Countries", *Biodiversity Conservation*, (1), pp. 34-50.

Barbier, E.B. (1994), "Valuing Environmental Functions: Tropical Wetlands", *Land Economics*, 70, pp. 155-173

Barbier, E.B. (2000), "Valuing the Environment as Input: Review of Applications to Mangrov-fishery linkages, *Ecological Economics*, Vol. 35(1), pp. 47-63.

Bateman, I.J. and K.G. Willis(eds.) (1999), *Valuing Environmental Resources: Theory and Practice of the Contingent Valuation Method in the US, EU, and Developing Countries*, New York: Oxford University Press.

Beioley, S. (1995), "Green Tourism—Soft or Sustainable?" *Insights*, May, pp. 75-89.

Bell, P.R.F. (1991), "Impact of Wastewater Discharges from Tourist Resorts on Eutrophicatin in Coral Reef Regions and Recommended Methods of Treatment", in M.L. Miller and J. Auyong (eds.), *Proceedings of the 1990 Congress on Coastal and Marine Tourism Honolulu, Hawaii*, Newport: National Coastal Resources Research and Development Institute.

Birundha, V.D. (2003), "Ecotourism—An Evaluation" in V.D. Birundha (ed.), *Environmental Changes Towards Tourism*, Kaniskha Publishers, New Delhi.

Blamey, E. (1996), "Ecotourism in the Third World", in E. Cater and G. Lawman (eds.) *Ecotourism: A Sustainable Option*? Chichester: John Wiley.

Blundell, R. (1991), "Consumer Behaviour: Theory and Empirical Evidence—A Survey", in A.J. Osuald (ed.), *Surveys in Economics*, Vol. 2, Oxford: Basil Blackwell.

Board, J., Sinclair, M.T. and C.M.S. Sutcliffe (1987), "A Protfolio Approach to Regional Tourism", *Built Environment*, Vol. 13, No. 2, pp. 124-137.

Bockstael, P.P. Bishop, R.C. and N.W. Bouwes (1991), "The Travel Cost Model for Lake Recreation: A Comparison of Two Methods for Incorporating Site Quality and Substitution Effects", *American Journal of Agricultural Economics*, 68(2), pp. 291-297.

Boissevain, J. (1977), *Tourism and Development in Malta*, Oxford University Press, New York.

Boo, E. (1990), *Ecotourism: The Potential and Pitfalls*, Washington: World Wildlife Fund.

Bote Gomez, V. (1993), "La necessaria revalorizaction de la actividad turistica espancta en una econimia terciarizade e integrade in la CEE", *Estudious Turisticos*, No. 118, pp. 5-26.

Boulding, R. (1985), "Planificacion del espaio turistico", *Editorial Trillas*, Maxico, pp. 9-20.

Broomfield, J.G. (1991), *Demand for Tourism in Fiji*, M.A. Dissertation in Development Economics, University of Kent at Canterbury.

Brown, G. Jr. and W. Henry (1989), *The Economic Value of Elephants*, London Environmental Economics Centre Paper 89-12, University College London.

Brown, W.G. and F. Nawas (1973), "Impact of Aggregation on the Estimation of Outdoor Recreation Demand Functions", *American Journal of Agricultural Economics*, 55: p. 246-259.

Bryden, J.M. (1973), *Tourism and Development: A Case Study of the Common Wealth Caribbean*, Cambridge University Press.

Buckley, P.J. (1993), "Tourism and Foreign Currency Receipts", *Annals of Tourism Research*, Vol. 20(2) pp. 361-367.

Cater, E. and Goodall, B. (1992), "Must Tourism Destroy Its Resources Base? In A.M. Mannin and S.R. Bowlty (eds.), *Environmental Issues in the 1990s*, Chichester, Wiley.

Ceballos, L.H. (1993), *Ecotourism: A Guide for Planners and Managers*, The Ecotourism Society, North Bennington, Vermount: 1-7.

Cesario, F.J. (1978), "Value of Time in Recreation Benefit Studies", *Land Economics*, 52(1), pp. 32-41.

Champ. R. T. Flares, N.E., Brown, M.F. and N. Chivers (2000), *Contingent Valuation: Controversies and Evidence*, San Diego: Department of Economics, University of California.

Chant, S. (1997), "Gender and Tourism Employment in Mexico and the Philippines", in M.T. Sinclair (ed.) *Gender, Work and Tourism*, London and New York: Routledge.

Chase, L., D. Lee, W. Schutzae and D. Anderson (1998), "Ecotourism Demand and Dr. Aerential Pricing of National Park Access in Costa Rica", *Land Economis*, 74 (4): pp. 466-482.

Chauhan, P. (2002), "Capturing Ecotourism Potential at Chilika Lake", Paper Presented in *International Workshop in Restoration of Chilika Lagoon*, CDA, Bhubaneswar, pp. 39-40.

Choe, K., Whittington, D. and D.T. Lauria (1996), "The Economic Benefits of Surface Water Quality Improvement in Developing Countries: A Case Study of Davas, Philippines", *Land Economics*, 72(4): pp. 519-527.

Chopra, K and S.K. Adhikari (2004), "Environment Development Linkages: Modelling a Wetland System for Ecological and Economic Value", *Environment and Development Economics*, Vol. 9. pp. 19-45.

Chopra, K. and G.K. Kadekodi (1999), "Valuation of Forest Resources: An Application of the Contingent Valuation Method", *Working Paper Series E/205/99*. Institute of Economic Growth, Delhi.

Choudhury, A.K., Sahu, N.C. and C. Parrings (2005), "Sustainable Management of Chilika Prawn Fisheries", in N. Sengupta and J. Bandyapadhyay (eds.), *Biodiversity and Quality of Life*, Macmillan India Ltd, New Delhi, pp. 253-272.

Choudhury, B.C. (2002) "Community based ecotourism in Chilika-Rushikulya-Gopalpur Ecoregion—A Proposal for Enhancing Livelihood Option and Restoration of Coastal Wetland Habitats in Orissa, Paper Presented in *International Workshop on Restoration of Chilika Lagoon*, CDA, Bhubaneswar, p. 45.

Clawson, M. (1959), *Methods of Measuring the Demand for and Value of Outdoor Recreation*, Reprint No. 10, Washington: Resource for the Future.

Clawson, M. and J.L. Knetsch (1966), *Economics of Outdoor Recreation*, Washington, D.C. John Hopkins University Press.

Conlin, M.V. and T. Baum (1995), *Island Tourism*, Chichester, John Wiley.

Cook, S.D., Steward, E. and K. Repass (1992), *Discover America: Tourism and the Environment*, Washington: Travel Industry Association of America/London: Belhaven.

Cukier, J., Norries, J. and G. Will (1996), "The Involvement of Women in the Tourism Industry of Bali, Indonesia", *Journal of Development Studies*, Vol. 33(2), pp. 248-270.

Curry, S. (1992), "Economic Adjustment Policies and the Hotel Sector in Jamaica", in P. Johnson and B. Thomas (eds.), *Perspectives on Tourism Policy*, London: Mansell.

Das, D.K. (2002), "Ned for Eco-tourism Development in Chilika Lagoon. An Equiry into its Growth, Problems and Strategies, Abstract, *International Workshop on Restoration of Chilika lagoon*, CDA, Bhubaneswar.

de Kadt, E. (1979), *Tourism Passport to Development*, Oxford: Oxford University Press.

Deaton, A.S. (1992), *Understanding Consumption*, Oxford: Clarendon Press.

Deaton, A.S. and J. Muellbauer (1980)a, "An Almost Ideal Demand System", *American Economic Review*, Vol. 70, No. 3, pp. 312-326.

Deaton, A.S. and J. Muellbauer (1980)b, *Economics and Consumer Behaviour*, Cambridge: Cambridge University Press.

Dekalt, E. (1979), *Tourism: Passport to Development? Perspective on the Social and Cultural Effects of Tourism in Developing Countries.* Oxford University Press, New York.

Dieke, P.V.C. (1993), "Tourism in the Zambia: Some Issues in Development Policy: *World Development*, Vol. 21(2), pp. 277-287.

Eagles, S. (1992), "Green Tourism-Soft or Sustainable?" *Insights,* May pp. 75-89.

Eber, S. (ed.) (1992), *Beyond the Green Horizon Principles for Sustainable Tourism,* Godabming World Wide Fund for Nature.

Economic and Social Commission for Asia and the Pacific (ESCAP) (1991). *Investment and Economic Cooperation in the Tourism Sector in Developing Asian Countries,* ESCAP Tourism Review. No. 8, Bangkok.

Ehrlich, P. and A. Ehrlich (1981), *Extinction,* Ballantine Books, New York.

Elkon, W (1975), "The Relation Between Tourism and Employment in Kenya and Tanzania", *Journal of Development Studies,* Vol. 11(2), pp. 123-130.

English, E.P. (1986), *The Great Escape? An Examination of North—South Tourism,* Ottawa: North-South Institute.

Euromonitor (1997), *World Tourism 1997,* London: Euromonitor.

EXA International/CHL, Consulting Group (1993), *The Economic Significance of Tourism in Zimbabwe,* EXA International.

Falian, P. and B. Kareher (1997), "The Impact of Aviation Upon the Atmosphere, An Assessment of Present Knowledge, Unconstrainties and Research Needs", *Earth,* Vol. 22, pp. 503-598.

Fankhausen, S. (1994), "The Social Costs of Greenhouse Gas Commissions: An Expected Value Approach", *Energy Journal,* Vol.15, No. 2. pp. 157-184.

Farber, S. (1988), "The Value of Coastal Wetlands for Recreation: An Application of Travel Cost and Contingent Valuation Methodologies", *Journal of Environmental Management*, 26, pp. 299-312.

Farrell, B.H. and D. Runyan (1991), "Ecology and Tourism", *Annals of Tourism Research*, Vol. 18, No. 1 pp. 26-40.

Fletcher, J.E. and B.H. Archer (1991), "The Development and Application of Multiplier Analysis", in C.P. Cooper (ed.), *Progress in Tourism, Recreation and Hospitality Management*, Vol. 1, London: Belhaven.

Forsyth, T. (1995)a, "Business Attitudes to Sustainable Tourism: Responsibility and Self Regulation in the UK Outgoing Tourism Industry", Paper Presented at the Sustainable Tourism World Conference, Lanzarate.

Forsyth, T. (1995)b, "Tourism and Agricultural Development in Thailand", *Annals of Tourism Research*, Vol. 22(4), pp. 877-900.

Freeman, A.N. (1993), *The Measurement of Environmental and Resource Values: Theory and Methods*, Washington, D.C. Resources for the Future.

Friedman, M. (1957), *A Theory of the Consumption Function*, Princeton, NJ: Princeton University Press.

Fujii, E., Khaled, M. and J. Mak (1987), "An Empirical Comparison of Systems of Demand Equations: An Application to Visitor Expenditure in Resort Destinations", *Philippine Review of Business and* Economics, Vol. 24, Nos. 1-2, pp. 79-102.

Garrod, G and K.G. Willis (1999), "Contingent Valuation Method", *Economic Valuation for the Environment*, Edward Elgar, U.K.

Garrod, G. and K. Willis. (1991), "The Environmental Economics Impact of Woodland: A Two-stage Hidonic Price of the Amenity Value of Forestry in Britain". *Applied Economics*, No. 24, pp. 715-728.

Glamecchini, S.G. (1993), "Sustainable Competitive Advantage in Service Industry, *Journal of Marketing*, Vol. 57, No. 4, pp. 83-99.

Gossling, S. (1999), "Eco tourism: a means to safeguard biodiversity and ecosystem functions?", *Ecological Economics*, Vol. 29, No. 2, pp. 303-320.

Government of Orissa (2003), Statistical Bulletin, Department of Tourism and Culture.

Government of Orissa (2004), Statistical Bulletin, Department of Tourism and Culture.

Grandstaff, S. and J.A. Dixon (1986), "Evaluation of Lumpinee Park in Bangkok, Thailand", in J.A. Dixon and M.M. Hufsc hmidt (eds.), *Economic Valuation Techniques for the Environment: A Case Study Workbook*, Baltimore: Johns Hopkins University Press.

Gray, H.P. (1966), "The Demand for International Travel by the United States and Canada", *International Economic Review*, Vol., No. 1, pp. 83-92.

Greene, W.H. (2000), *Econometric Analysis*, Fourth Edition, New Jersey, Prentice Hall.

Grilicher, Z. (1971), *Price Indexes and Quality Change*, Harvard University Press, Cambridge, Massachusetls.

Guandhi, H. and C.K. Boey (1986), "Demand Elasticities of Tourism in Singapore", *Tourism Management*, Vol. 7, No. 4, pp. 239-253.

Gujarati, D.N. (2004), *Basic Econometrics*, McGraw Hill, Singapore, 4th Edition.

Gun, R.L. and W.E. Martin (1974), "Problems and Solutions in Estimating the Demand for and Value of Rural Outdoor Recreation". *American Journal of Agricultural Economics*, 56: p. 558-566.

Hadker, N., Sharma, S., David, A. and T.R. Maraleedharam (1997), "Willingness–to-pay for Borivli National Park: Evidence from Contingent Valuation", *Ecological Economics* Vol. 21, pp. 105-122.

Hanley, N. and R. Ruffell (1993), "The Valuation of Forest Characteristics", *Queen's Discussion Paper* 849.

Hanley, N. Shogren, J.F., and B. White (1997), *Environmental Economics in Theory and Practice,* Macmillan Press Ltd. UK.

Hawkins, J.P. and C.M. Roberts, (1994), "The Growth of Coastal Tourism in the Red Sea: Present and Future Effects on Coral Reefs", *Ambio,* No. 23, pp. 503-508.

Helu Thaman, K. (1992), "Beyond Hula, Hotels and Handicrafts", *In Focus,* Vol. 4, Summer, pp. 8-9.

Holloway, J.C. (1994), *The Business of Tourism,* Pitman, London.

Hotelling, H. (1947), *"The Economics of Public Recreation: The Prewill Report"*, Washington, D.C. National Park Services.

Huband, M. (1997), "Something Big to Sing About", *Financial Times,* 13 May.

IUCN (International Union for the Conservation of Nature) (1992), World Heritage Twenty Year Later, IUCN, Gland, Switzerland, Compiled by J. Jhorsell.

Jenner, P. and C. Smith (1992), "The Tourism Industry and the Environment", Special Report 2453, London: Economist Intelligence Unit.

Johnson, P. and J. Ashworth (1990), "Modelling Tourism Demand: A Summary Review", *Leisure Studies,* Vol. 9. No. 2, pp. 145-160.

Jud, G.D. and H. Joseph (1974), "International Demand for Latin American Tourism", *Growth and Change,* No. 5, pp. 25-31.

Kadekodi, G.K. and A. Nayampalli (2005), "Reversing Biodiversity Degradation: A Case of Chilika Lake", in N.C. Sahu and A.K. Choudhury (eds). *Dimensions of Environmental and Ecological Economics,* Universities Press, Hyderabad.

Kaosaard, M.D. Patmasiriwat, T. Panayotou, and J.R. Deshazo (1995), *Green Financing Valuation and Financing of Khas Yai National Park in Thailand,* Thailand Development Research Institute, Bangkok.

Katchen, M. J. a nd S.D. Reiling (2000), "Environmental Attitudes, Motivations and Contingent Valuation of Non use Values: A Case Study Involving Endangered Species", *Ecological Economics.* 32(1), pp. 93-107.

Kentsch, J.L. (1963), "Outdoor Recreation Demand and Values", *Land Economics*, 39, pp. 387-96.

Knudsen O. and A. Parnes (1975), *Trade Instability and Economic Development*, Lenington, MA: D.C. Heath.

Krippendorf, J. (1987), *The Holiday Makers,* London, Heinmann.

Kuss, F.R. Grafe, A.R. and J.J. Vaske (1990), "Recreation impact and carrying capacity", *National Parks and Conservation Association,* Vol. 1 &2, Washington D.C.

Lang, S.D., Steward, E. and K.Repass (1992), *Discover America; Tourism and the Environment,* Washington: Travel Industry Association of America/London: Belhamen.

Lawton, J.H. and R.M. Hay (1995), *Extinction Rates,* Oxford University Press, New York.

Lea, J. (1988), *Tourism and Development in the Third World,* London and New York: Routledge.

Lee, G. (1987), "Tourism as a Factor in Development Cooperation", *Tourism Management*, Vol. 8(1), pp. 2-19.

Lee, W. (1991), "Prostitution and Tourism in South-East Asia", N. Reclift and M.T. Sinclair (eds.), *Working Women: International Perspectives on Labour and Gender Ideology,* London and New York: Routledge.

Lenak, S.K. (2002), "Development Community Driven Eco-tourism Strategy: An Analysis of Orientation Programme on Ecotourism for Boatman at Satapada", Abstract, *International Workshop on Restoration of Chilika Lagoon,* CDA, Bhubaneswar, pp. 38.

Lenka, S.K. (1998), "An Approach to Community Based Ecotourism Planning for Chilika lagoon", *Proceeding of the Workshop on Sustainable Development of Chilika Lagoon,* CDA, Bhubaneswar: pp. 318-323.

Levy, D.E. and P.B. Lerch (1991), "Tourism as a Factor in Developing: Implications for Gender and Work in Barbados", *Gender and Society,* Vol. 5(1), pp. 67-85.

Li, C.Z. (1994), "Semi-Parametric Estimation of the Binary Choice Model for Contingent Valuation", *Land Economics*, 72(4), pp. 462-472.

Lockwood, M., Loomis, J. and T. Delacy (1993), "A Contingent Valuation Survey and Benefit, Cost Analysis of Forest Preservation in East Gippsland. Australia, *Journal of Environmental Management* No. 38, pp. 233-243.

Long, V.H. (1991), "Government-Industry-Community Interaction in Tourism Development in Mexico", in M.T. Sinclair and M.J. Stabler (eds.). *The Tourism Industry: An International Analysis*, Wallingford: CAB International.

Long, V.H. and S.L. Kindon (1997), "Gender and Tourism Development in Balinese villages", in M.T. Sinclair (eds.), *Gender, Work and Tourism*, London and New York: Routledge.

Loomis, J.B. and Caban, A.G. (1998), "A Willingness to Pay Function for Protecting Acres of Spotled owl Habitat from Fire", *Ecological Economics*, 25(3), pp. 315-322.

Loomis, J.B., Creel, M. and T. Park 1991, "Comparing Benefit Estimates from Travel Cost and Contingent Valuation Using Conference Intervals from Hicksion Welfare Measures", *Applied Economics*, No. 23, pp. 1725-1731.

Loonis, J.B, Brown, T. Cucero, B and G. Peterson (1991), "Improving Validity Experiments of Contingent Valuation Methods. Results of Efforts to Reduce the Disparity of Hypothetical and Actual Willingness to Pay", *Land Economics*, 72(4), pp. 450-461.

Markowitz, H. (1959), *Portfolio Selection: Efficient Diversification of Investment*, New York: John Wiley.

Mc Kean, P.F. (1997), "Towards a Theoretical Analysis of Tourism: Economic Dualism and Cultural Involution in Bali", *Hosts and Guests: The Anthropology of Tourism*. Edited by Volume L. Smith, University of Pennsylvania, Philadelphia.

McConnell, K.E. (1975), "Some Problems in Estimating the Demand for Outdoor Recreation", *American Journal of Agricultural Economics*. 78(1) pp. 330-334.

McConnell, K.E. (1992), "On-site Time in The Demand for Recreation", American *Journal of Agricultural Economics*, 74, pp. 918-925.

McFadden, P.H. (1977), "A New Approach to the Evaluation of Non-priced Recreational Resources", *Land Economics*, 11 (2), 135-151.

Mckean, J.R., Johnson, D.M. and R.G. Walsh (1995), "Valuing Time in Travel Cost Demand Analysis: An Empirical Investigation", *Land Economics*, 71(1): pp. 96-105.

Milne, S.S. (1987), "Differential Multipliers", *Annals of Tourism Research*, Vol. 14(4), pp. 499-515.

Mohanty, S. (1999), "Socio-Economic Study of Peripheral Village of Chilika Lake in Relation to its Ecological Changes during the Post Independence Period". Thesis Submitted Doctor of Philosophy in History, P.G. Department of History, Berhampur University, Orissa.

Mudambi. R. (1994), "A Ricardian Excursion to Bermuda: An Estimation of Mixed Startegy Equilibrium", *Applied Economics*, No. 26, pp. 927-936.

Murthy, M.N. (2000), "Valuation of Water", Institute of Economic Growth University Enclave, Delhi, India.

Nanda, B.K. (2002), "Environmental Education and Awareness Promotion on Eco-tourism and Community Development with Reference to Chilika Lagoon, Orissa, Abstract, *International Workshop on Restoration of Chilika Lagoon*, CDA, Bhubaneswar, pp. 43.

Nash, D. (1981), "Tourism as an Anthropological Subject", *Current Anthropology*, Gitanjali Publishing House, New Delhi.

O' Hagan, J.W. and M.J. Harrison (1984), "Market Shares of US Tourist Expenditure in Europe: An Econometric Analysis", *Applied Economics*, Vol. 16, No. 6, pp. 919-931.

OTDC Report (2000), *Orissa Tourism Accommodation Facilities and Package Tours*, Department of Tourism, Goo.

Parida, B.B. (2002), "Community Based Tourism in Chilika", Abstract, *International Workshop on Restoration of Chilika Lagoon*, CDA, Bhubaneswar, p. 43.

Pate, J. and J.B. Loomis (1997), "The Effect of Distance on Willingness to Pay Values: A Case Study of Wetlands and Salman in California", *Ecological Economics*, Vol. 20. pp. 199-207.

Pawson, I.G., Stanford, D.D., Adams, V.A. and M. Nurbu (1984), "Growth of Tourism in Nepal's Everest Region: Impact on the Physical Environment and Structure of Human Settlements", *Mountain Research and Development*, Vol. 4, No. 3, pp. 237-246.

Pearce, D. (1985), "Tourism and Environmental Research: A Review, "*International Journal of Environmental Studies*, Vol. 25, No. 4, pp. 247-255.

Pearce, D.W. and Moran, D. (1994), *The Economic Value of Biodiversity*, IUCN —The World Conservation Union, Earth-scan, London.

Pearce, D.W., Markandya, A. and E.B. Barbier (1991), "Valuing the Environment", *Blueprint for a Green Economy*, pp. 51-81.

Pindyck, R.S. and Rubinfeld, D.L. (2003 & 2004), *Econometric Models and Economic Forecasts*, Fifth & Sixth Edition, McGraw—Hill Book Co-Singapore.

Pyo, S.S., Uysal M. and R.W. McLellan (1991), "A Linear Expenditure Model for Tourism Demand", *Annals of Tourism Research*, No. 18, pp. 443-454.

Randall, A. and J. Stall (1988), "Existence Values and in a Total Valuation Framework", in R.D. Row and I.G. Chestnut (eds.), *Managing Air Quality and Science Resources at National Parks and Wilderness Areas*, West-view Press, Boulder Co.

Rao, A. (1986), *Tourism and Export Instability in Fiji*, Occasional Papers in Economics Development No. 2, Faculty of Economic Studies, University of New England, Australia.

Rasen. S. (1974), "Hedonic Prices and Implicit Market: Production Differentiation in Pure Competition", *Journal of Political Economy*, Vol. 82, No. 1, pp. 34-55.

Rath, A. (2000), *Preservation Value of a Wetland Ecosystem: A Case Study of Chilika*, A Dissertation Submitted to the University of Delhi for Degree of Master of Philosophy.

Romeril, M. (1989), "Tourism and Environment Accord or Discord",*Tourism Management*, Vol. 10, No. 3, pp. 204-208.

Romeril, M. (1989), "Tourism and the Environment: Accord or Discord", *Tourism Management*, Vol. 10, No. 3, pp. 204-208.

Rosen, S. (1974), "Hedonistic Price and Implicit Markets: Product Differentiation in Pure Competition", *Journal of Political Economy*, 82, pp. 34-55.

Rout, S. (2003), *Option Value of Chandaka Forest in Orissa*, Dissertation Submitted to the Master of Philosophy in Economics, P.G. Department of Economics, Berhampur University, Orissa.

Sadler, P., archer, B.H. and C. Owen (1973), *Regional Income Multipliers*, Occasional Papers in Economics, No. 1, Bangor: University of Wales Press.

Sakai, M.Y. (1988), "A Micro-Analysis of Business Travel Demand", *Applied Economics*, No. 20, pp. 1481-1491.

Samarasuriya, S. (1982), Who Needs Tourism? Employment for Women in the Holiday Industry of Sudugama, Sri Lanka, Colombo, Research Project for Women and Development.

Sathiendra Kumar, R. and C. Tisdell (1989), "Tourism and the Economic Development of Maldives", *Annals of Tourism Research*, Vol. 16(2), pp. 245-249.

Scott, J. (1997), "Chances and Choice: Women and Tourism in Northern Cyprus", in M.T. Sinclair, (eds.), *Gender, Work and Tourism*, London and New York: Routledge.

Seeberger, V.M. (1992), *Potential European Demand for Tourism in Reunion Island*, M.A. Dissertation in Development Economics, University of Kend at Canterbury.

Sekhsaria, P. (2002), "The Ecotourism Judgement", *The Hindu Mangazine*, Nov. 3rd, 1.

Seller, C., Stoll, J.R. and J.P. Chavas (1985), "Validation of Empirical Measures of Welfare Change: A Comparison of Non-market Technique", *Land Economics*, 61(2), pp. 156-176.

Shamsuddin, S. (1995), *Tourism Demand in Peninsular Malaysia*, M.A. Dissertation in Development Economics, University of Kent at Canterbury.

Shaw, W.D. (1992), "Searching for the Opportunity Cost of An Individual's Time", *Land Economics*, 68, pp. 107-115.

Sheldon, P.J. (1990) "A Review of Tourism Expenditure Research", in C.P. Cooper (ed.), *Progress in Tourism, Recreation and Hospitality Management*, Volume Two, London: Belhaven.

Sheldon, P.J. (1990), "A Review of Tourism Expenditure Research", in C.P. Cooper (ed.), *Progress in Tourism, Recreation and Hospitality Management*, Volume Two, London: Belhaven.

Sinclair, M.T. (1990), *Tourism Development of Kenya*, Washington, D.C.: World Bank.

Sinclair, M.T. (1991)a, "The Economics of Tourism", in C.P. Cooper (ed.). *Progress in Tourism, Recreation and Hospitality Management*, Volume Two, London: Belhaven.

Sinclair, M.T. (1991)b, "The Tourism Industry and Foreign Exchange Leakages in Developing Country", in M.T. Sinclair and J.J. Stables (eds.), *The Tourism Industry: An International Analysis*, Wallingford; CAB International.

Sinclair, M.T. (1992), "Tour Operators and Tourism Development Policies in Kenya", *Annals of Tourism Research*, Vol. 19(3), pp. 555-558.

Sinclair, M.T. (1997), "Issues and Theories of Gender and Work in Tourism", in M.T. Sinclair (eds.), *Gender, Work and Tourism*, London and New York: Routledge.

Sinclair, M.T. and A. Tsegaye (1990), "International Tourism and Export Instability", *Journal of Development Studies*, Vol. 26, No. 3, pp. 487-504.

Sinclair, M.T. and A. Tsegaye (1990), "International Tourism and Export Instability", *Journal of Development of Studies*, Vol. 26(3), pp. 487-505.

Sinclair, M.T. and C.M.S. Sutchiffe (1978), "The First Round of the Keynesian Income Multiplies", *Scottish Journal of Political Economy*, Vol. 25(2), pp. 177-186.

Sinclair, M.T. and C.M.S. Sutchiffe (1988), "The Estimation of Keynesian Income Multipliers at the Sub-National Level", *Applied Economics*, 20(11), pp. 1435-1444.

Sinclair, M.T. and C.M.S. Sutchiffe (1989), "Truncated Income Multipliers and Local Income Generation Over Time", *Applied Economics*, Vol. 21(2), pp. 1621-1630.

Sinclair, M.T. and M.J. Stabler (1997), *The Economics of Tourism*, London and New York: Routledge.

Smeral, E. (1988), "Tourism Demand, Economic Theory and Econometrics: An Integrated Approach", *Journal of Travel Research*, Vol. 26, No. 4, pp. 38-43.

Smith V.K. Palmquist, R.B. and P. Jakus (1991), "Combining Farrel Frontier and Hedonic Travel Cost Models for Valuing Estuarine Quality", *Review of Economics and Statistics*, Vol. 63, No. 4, pp. 694-699.

Smith, C. and P. Jenaer (1994) "Travel Agents in Europe", *Travel and Tourism Analyst*, No. 3, pp. 56-72.

Stabler, M.J. and B. Gooddall, (1996), "Environmental Auditing in Planning for Sustainable Island Tourism", in L. Brigudlio, B. Archer, J. Jafari and G. Wall, *Sustainable Tourism in Islands and Small Stales: Issues and Policies*, London: Pinter (Cassell).

Stronge, W.B. and M. Redman (1982), "US Tourism in Mexico: An Empirical Analysis", *Annals of Tourism Research*, Vol. 9. No. 1, pp. 21-35.

Sutherland, R.J. and R.G. Walsh (1985), "Effect of Distance on the Preservation Value of Water Quality", *Land Economics*, 61(3), pp. 281-291.

Svedsater, H. (2003), "Economic Valuation of the Environment: How Citizens Make Sense of Contingent Valuation Questions" *Land Economics,* Vol. 79, pp. 122-135.

Swain, S.K. (2002), "Formulation Conservation Strategy for the Development of Ecotourism: A Study of the Chilika Lake", Abstract, *International Workshop on Restoration of Chilika Lagoon,* CDA, Bhubaneswar, p. 45.

Syndyo, D.M. and F.N. Pertet (1984), "Tourism and Its Impact on Wildlife in Kenya", *Industry and Environment,* Vol. 7(1), pp. 14-19.

Syriopoulos, T. (1995), "A Dynamic Model of Demand for Mediterranean Tourism", *International Review of Applied Economics,* Vol. 9, No. 3, pp. 318-336.

Syriopoulos, T. and M.T. Sinclair (1993), "An Econometric Study of Tourism in Mediterranean Countries, *Applied Economics,* Vol. 25, No. 12, pp. 1541-1552.

Tantrigama, G. (1998), "Carrying Capacity and Sustainability of Coastal Tourism in Hikkaduwa, Sri Lanka and Calangute, Goa, India", *Tropical Coasts,* Vol. 5(1), pp. 16-20.

Tarai, D.K. (2003), "Travel Cost Demand Function for Recreation in Chilika Lake". Dissertation Submitted for Master of Philosophy in Economics, P.G. Department of Economics, Berhampur University, Orissa.

Thirlwall, A.P. (1976), "The Balance of Payments Constraint as an Explanation of International Growth Rate Differences", *Banca Nazionale de Lovoro Quartely Review,* No. 128, pp. 45-53.

Thirlwall, A.P. (1986), "A General Model of Growth and Development on Kaldorian Lines", *Oxford Economic Papers,* No. 38, pp. 199-219.

Thirlwall, A.P. and M. Nureldin – Hussein (1982), "The Balance of Payments Constraint, Capital Flows and Growth Rate Differences Between Developing Countries", *Oxford Economic Papers,* No. 34, pp. 498-510.

Tourism Concern (1995), "Our Holiday, Thesis Homes" (Special Issue on People Displaced by Tourism), *In Focus,* Vol. 15. Spring, pp. 3-13.

Trisal, C.L., Chauhan, M. (1998), *Chilika Lake, a Guidelines for Ecotourism Development*, Wetlands International South Asia, New Delhi.

United Nations Conference on Trade and Development (UNCTAD) (1973), *Element of Tourism Policy in Developing Countries*, Report by the Secretariat of UNCTAD, TD/B/C. 3/89, Add. 3, Geneva: UNCTAD.

Uysal, M. and J.L. Crompton (1984), "Determinants of Demand for International Tourism Flows in Turkey", *Tourism Management*, Vol. 5, No. 4, pp. 288-297.

Varley, R.C.G. (1978), *Tourism in Fiji: Some Economic and Social Problems*, Occasional Papers in Economics, No. 12, University of Wales Press Bangor.

W.T.O. (World Tourism Organisation), (1992), Guideline: Development of National Parks and Protected Areas for Tourism, Madrid: pp. 1-20.

Walsh, R.G., Johnson, D.M., and J.R. McKean (1992), "Benefit Transfer of Outdoor Recreation Demand Studies, 1968-88", *Water Resources Research*, 28, pp. 707-713

Wanhill, S.R.C. (1988), "Tourism Multipliers under Capacity Constraints", *Service Industries Journals*, pp. 136-142.

Ward, W.D. (1992), "Measuring the Cost of Time in Recreation Demand Analysis: Comment", American Journal of Agricultural Economics, 65, pp. 167-168.

Welson, E.U. (1985), "The Biological Diversity Crisis", *Bioscience* Vol. 35, No. 1, pp. 700-706.

White, K.J. (1982), "The Demand for International Travel: A System-wide Analysis for U.S. Travel to Western Europe", *Discussion Paper No.* 82-28, University of British Columbia, Canada.

Wight, P. (1993), "Ecotourism: Ethics or Eco-sell?" *Journal of Travel Research*, No. 31, pp. 35-47.

Wight, P. (1994), "The Greening of the Hospitality Industry: Economic and Environmental in P.K. Sense and V. Seaton (ed.), Tourism: *The State of the Art*, Chichester, Wiley.

Willis, K and G. Garrod (1993), "The Vale of Waterside Properties: Estimating the Impact of Waterway and Canals on Property Value Through Hedonic Price Models and Contingent Valuation Methods", *Countryside Change Unit Working Paper* 44, Neweastle: University of Newcastle.

Willis, K. and G. Garrod (1993)b, "Valuing Wildlife: The Benefits of Wildlife Trust", *Countryside Change Unit Working Paper.* 46, Newcastle: University of Newcastle.

Yaping, Du (1998), *The Value of Improved Water Quality for Recreation in East Lake, Wuhan, Chira: Application of Contingent Valuation and Travel Cost Methods,* EEPSEA Research Report Series, Economy and Environment Programme for Southeast Asia, Tanglin, Singapore.

Index

V

W

Z